Tucholsky Wagner Zola Scott Sydow Freud Schlegel
Turgenev Wallace Fonatne
Twain Walther von der Vogelweide Fouqué Friedrich II. von Preußen
Weber Freiligrath Frey
Fechner Fichte Weiße Rose von Fallersleben Kant Ernst Richthofen Frommel
Engels Fielding Hölderlin Tacitus Dumas
Fehrs Faber Flaubert Eichendorff
Eliasberg Ebner Eschenbach
Feuerbach Maximilian I. von Habsburg Fock Eliot Zweig
Ewald Vergil
Goethe Elisabeth von Österreich London
Mendelssohn Balzac Shakespeare Dostojewski Ganghofer
Trackl Lichtenberg Rathenau Doyle Gjellerup
Stevenson Hambruch
Mommsen Thoma Tolstoi Lenz Hanrieder Droste-Hülshoff
Dach Verne von Arnim Hägele Hauff Humboldt
Karrillon Reuter Rousseau Hagen Hauptmann Gautier
Garschin Defoe Baudelaire
Damaschke Descartes Hebbel
Hegel Kussmaul Herder
Wolfram von Eschenbach Dickens Schopenhauer Rilke George
Bronner Darwin Melville Grimm Jerome Bebel
Campe Horváth Aristoteles Federer Proust
Bismarck Vigny Barlach Voltaire Herodot
Gengenbach Heine
Storm Casanova Tersteegen Gilm Grillparzer Georgy
Chamberlain Lessing Langbein Gryphius
Brentano Lafontaine
Strachwitz Claudius Schiller Kralik Iffland Sokrates
Katharina II. von Rußland Bellamy Schilling
Gerstäcker Raabe Gibbon Tschechow
Löns Hesse Hoffmann Gogol Wilde Gleim Vulpius
Luther Heym Hofmannsthal Klee Hölty Morgenstern Goedicke
Roth Heyse Klopstock Kleist
Luxemburg Puschkin Homer Mörike Musil
La Roche Horaz
Machiavelli Kierkegaard Kraft Kraus
Navarra Aurel Musset Moltke
Nestroy Marie de France Lamprecht Kind Kirchhoff Hugo
Laotse Ipsen Liebknecht
Nietzsche Nansen
Marx Lassalle Gorki Klett Leibniz Ringelnatz
von Ossietzky May vom Stein Lawrence Irving
Petalozzi Platon Knigge
Pückler Michelangelo Kafka
Sachs Poe Kock
Liebermann Korolenko
de Sade Praetorius Mistral Zetkin

The publishing house tredition has created the series **TREDITION CLASSICS**. It contains classical literature works from over two thousand years. Most of these titles have been out of print and off the bookstore shelves for decades.

The book series is intended to preserve the cultural legacy and to promote the timeless works of classical literature. As a reader of a **TREDITION CLASSICS** book, the reader supports the mission to save many of the amazing works of world literature from oblivion.

The symbol of **TREDITION CLASSICS** is Johannes Gutenberg (1400 – 1468), the inventor of movable type printing.

With the series, tredition intends to make thousands of international literature classics available in printed format again – worldwide.

All books are available at book retailers worldwide in paperback and in hardcover. For more information please visit: www.tredition.com

tredition was established in 2006 by Sandra Latusseck and Soenke Schulz. Based in Hamburg, Germany, tredition offers publishing solutions to authors and publishing houses, combined with worldwide distribution of printed and digital book content. tredition is uniquely positioned to enable authors and publishing houses to create books on their own terms and without conventional manufacturing risks.

For more information please visit: www.tredition.com

Elements of Plumbing

Samuel Edward Dibble

Imprint

This book is part of the TREDITION CLASSICS series.

Author: Samuel Edward Dibble
Cover design: toepferschumann, Berlin (Germany)

Publisher: tradition GmbH, Hamburg (Germany)
ISBN: 978-3-8491-5274-1

www.tredition.com
www.tredition.de

Copyright:
The content of this book is sourced from the public domain.

The intention of the TREDITION CLASSICS series is to make world literature in the public domain available in printed format. Literary enthusiasts and organizations worldwide have scanned and digitally edited the original texts. tredition has subsequently formatted and redesigned the content into a modern reading layout. Therefore, we cannot guarantee the exact reproduction of the original format of a particular historic edition. Please also note that no modifications have been made to the spelling, therefore it may differ from the orthography used today.

PREFACE

In preparing this manuscript the author has had in mind the needs of young men having no technical instruction who are anxious to become proficient in the art of Plumbing. As a consequence each exercise is minutely described and illustrated; so much so, perhaps, that an experienced mechanic may find it too simple for skilled hands and a mature mind. But the beginner will not find the exercises too elaborately described and will profit by careful study. Years of experience and observation have shown the author that the methods herein described are entirely practical and are in common use today.

The various exercises in lead work will acquaint the beginner with the correct use of tools and metals. The exercises in iron pipe work have also been detailed to show the correct installation of jobs.

Together with the study of this book the subjects of Mathematics, Physics, Chemistry, Drawing and English should be taken. These subjects as they bear on Plumbing are invaluable to the mechanic in his future connection with the trade.

The author is indebted for the illustrations of fixtures in the chapter covering the development of plumbing fixtures, to the Thomas Maddock's Sons Co., Standard Sanitary Mfg. Co., and The Trenton Potteries Co.

Samuel Edward Dibble.
Pittsburgh, *December, 1917.*

[vi]

CONTENTS

Preface

Chapter

I.	Plumbing Fixtures and Trade
II.	The Use and Care of the Soldering Iron—Fluxes—Making Different Soldering Joints
III.	Mixtures of Solders for Soldering Iron and Wiping—Care of Solders—Melting Points of Metals and Alloys
IV.	Making and Caring of Wiping Cloths
V.	Preparing and Wiping Joints
VI.	Preparing and Wiping Joints (*Continued*)
VII.	Laying Terra-cotta and Making Connections to Public Sewers—Water Connections to Mains in Streets
VIII.	Installing of French or Sub-soil Drains
IX.	Storm and Sanitary Drainage with Sewage Disposal in View
X.	Soil and Waste Pipes and Vents—Tests
XI.	House Traps, Fresh-air Connections, Drum Traps, and Non-syphoning Traps
XII.	Pipe Threading
XIII.	Cold-water Supply—Test
XIV.	Hot-water Heaters—Instantaneous Coil and Storage Tanks—Return Circulation, Hot-water Lines and Expansion
XV.	Insulation of Piping to Eliminate Conduction, Radiation, Freezing and Noise
XVI.	"Durham" or "Screw Pipe" Work—Pipe and Fittings
XVII.	Gas Fittings, Pipe and Fittings, Threading, Measuring and Testing
XVIII.	Plumbing Codes

[1]

ELEMENTS OF PLUMBING

CHAPTER I

Plumbing Fixtures and Trade

Modern plumbing as a trade is the arranging and running of pipes to supply pure water to buildings, the erecting of fixtures for the use of this supply, and the installing of other pipes for the resulting waste water. The work of the trade divides itself therefore into two parts: first the providing an adequate supply of water; and second, the disposing of this water after use. The first division offers few problems to the plumber, little variety in the layout being possible, and the result depending mostly upon the arrangement of the pipes and fittings; but the second division calls for careful study in the arrangement, good workmanship in the installing, and individual attention to each fixture.

The trade had its beginnings in merely supplying fresh water to a community. This was done by means of trenching, or conveying water from lakes, rivers, or springs through wooden pipes or open troughs. By easy stages the trade improved and enlarged its scope, until at the present time it is able to provide for the adequate distribution of tons of water under high pressure furnished by the city water works.

In the early years of the trade the question of the disposal of the waste water was easily answered, for it was allowed to be discharged onto the ground to seek its own course. But with the increased amount of water available, the waste-water problem has enlarged until today it plays the most [2] important part of plumbing, and the trade has had to change to meet this waste-water problem.

The first simple system of a pipe running from the sink to a point outside the building was sufficient. As larger buildings came into use and communities were more thickly populated, the plumbing problem demanded thought and intense study. The waste pipes from fixtures had to be so arranged that it would be impossible for foul odors and germ-laden air to enter the building through a plumbing fixture. The importance of this is evidenced by the plumbing laws now in use throughout the country.

One of the first plumbing fixtures put into common use was a hollowed-out stone which served as a sink. It was with considerable interest that the writer saw a sink of this kind in actual use in the summer of 1915, at a house in a New England village. This sink had been in service for about 100 years. From this beginning the well-known fixtures of today have developed. The demand for moderate priced, sanitary closets, lavatories, and baths has led to the rapid improvement seen in plumbing fixtures. In the development of these fixtures, as soon as a bad feature was recognized the fixture was at once discarded, until now the market offers fixtures as mechanically fine as can be produced. Plumbing fixtures were at first manufactured so that it was necessary to support them on a wooden frame, and this frame was enclosed in wood. The enclosure made by this framework soon became foul and filthy and a breeding place for all kinds of disease germs and vermin. This bad feature was overcome by the introduction of open plumbing, that is, fixtures so made that the enclosure of wood could be done away with. The open plumbing allowed a free circulation of air around the fixture and exposed pipes, thereby making the outside of the fixture and its immediate surroundings free from all the bad features of the closed plumbing. Plenty of fresh air and plenty of [3] light are necessary for good sanitary plumbing.

Fig. 1.—Pan closet (English).

The materials of which the first open-plumbing fixtures were made consisted of marble, copper, zinc, slate, iron, and clay. Time soon proved that marble and slate were absorbent, copper and zinc soon leaked from wear, iron rusted, and clay cracked and lacked strength; therefore these materials soon became insanitary, and foul odors were easily detected rising from the fixture. Besides these materials being insanitary, the fact that a fixture was constructed using a number of sections proved that joints and seams were insanitary features on a fixture. For instance, in a marble lavatory constructed by using one piece for the top, another for the bowl, and still another for the back, filth accumulated at every joint and seam. Following this condition, developed the iron enameled and earthenware fixtures, constructed without seams and with a smooth, even, glossy white finish. The fact that these fixtures are made of material that is non-absorbent adds to their value as sanitary plumbing fixtures.

Another problem which is as important as the foregoing is the proper flushing, that is, the supplying of sufficient water in a manner designed to cleanse the fixture properly.

The development of sanitary earthenware illustrates how the above problems were satisfactorily solved. In the city of London a law compelling the use of drains was enforced, and in the early 70's the effect of this law was felt in this country. The introduction at this time of the mechanical water closet, known as the "pan closet," and the English plumbing material which was brought to this country [4] was the beginning of "American plumbing," which today outstrips that of any other country in the world. The "pan closet" continued in use for some time until the "valve closet" was introduced as a more sanitary fixture. Closely following these closets, in 1880, the plunger closet became popular as a still more sanitary fixture. The plunger closet continued in use until the present all-earthenware closet bowl drove all other makes from the market. The American development of the earthenware closet bowl put the American sanitary fixture far ahead of the English improvements, as the American earthenware is superior and the sanitary features of the bowls are nearer perfection.

Fig. 2.—Pan closet (American).

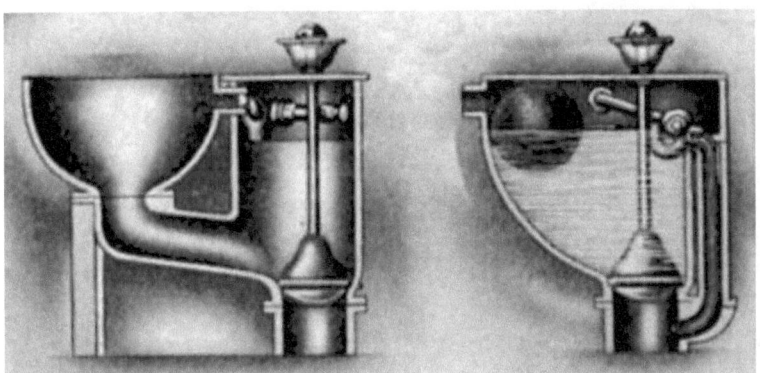

Fig. 3.—Plunger closet.

When the washout bowl was introduced it was considered perfection. The hopper closet bowl, which was nothing [5] more than a funnel-shaped bowl placed on top of a trap, was placed in competition with the washout bowl. There are a number of these bowls now in use and also being manufactured. However, large cities prohibit their use.

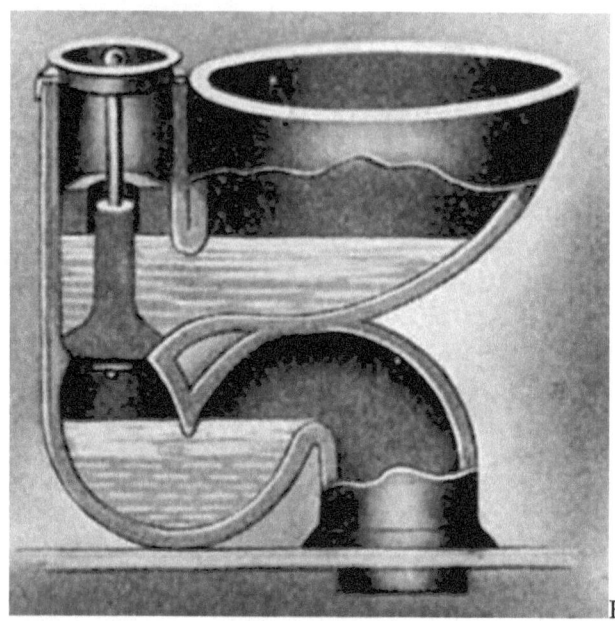

Fig. 4. — Plunger closet.

To quote Thomas Maddock's Sons Co.: "In 1876 Wm. Smith of San Francisco patented a water closet which employed a jet to assist in emptying the bowl and the development of this principle is due entirely to the potter, who had gradually and by costly experiment become the determining factor in the evolution of the water closet." With this improvement it became possible to do away with the boxing-in of the bowl which up to this time had been necessary. Closet bowls of today are made of vitreous body which does not permit crazing or discoloring of the ware. A study of the illustrations which show the evolution of the closet bowl should be of interest to the student as well as to the apprentice and journeyman. The bath tub developed from a gouged-out stone, in which water could be stored and used for bathing purposes, to our present-day enameled iron and earthenware tubs. The development did not progress very rapidly until about 25 years ago. Since then every feature of the tub has been improved, and from a sanitary [6] standpoint the tubs of today cannot be improved. The bath tub has become an American custom, as the people in this country have demanded that they have sanitary equipment in their homes, while in the European countries this demand has not developed.

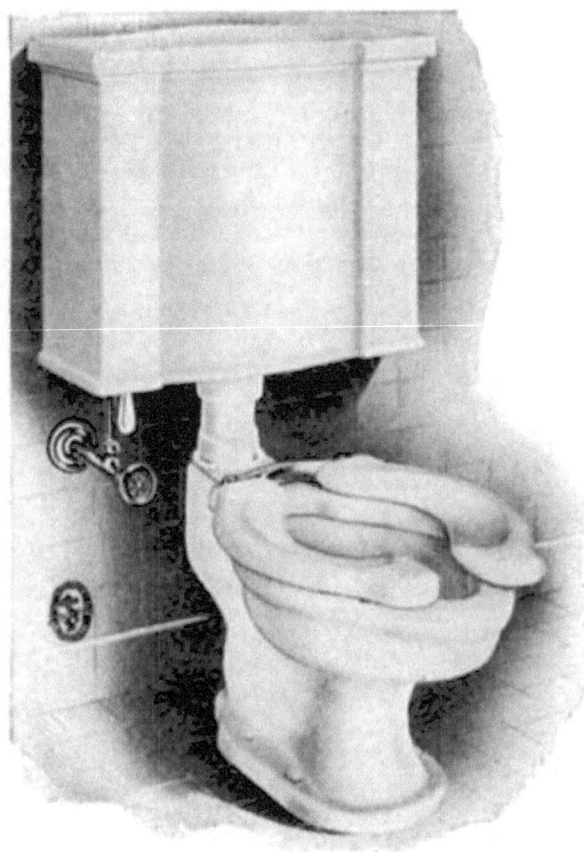

Fig. 5.— Modern low-tank closet.

The first tubs used in this country were of wood lined with copper or zinc, and were built in or boxed in with wood panelling. The plumbing ordinances of today prohibit this boxing as it proved to be a breeding place for vermin, etc. As the illustration shows, the woodwork encasing the tub was in a great many cases beautifully carved and finished.

The placing on the market of a steel-clad tub, a steel tub with a copper lining, which did away with the boxing, was a big improvement as far as sanitary reasons were [7] concerned as well as a reduction in cost of tubs. These tubs were set up on legs which permitted cleaning and provided good ventilation all around. With these features they drove all other tubs from the market. The copper and zinc were found to be hard to keep clean and they were soon replaced by the iron enamelled and earthenware tubs. The finish on these tubs being white and non-absorbent makes them highly acceptable as sanitary fixtures. A study of the illustrations will show how progress has been made in design as well as in sanitary features.

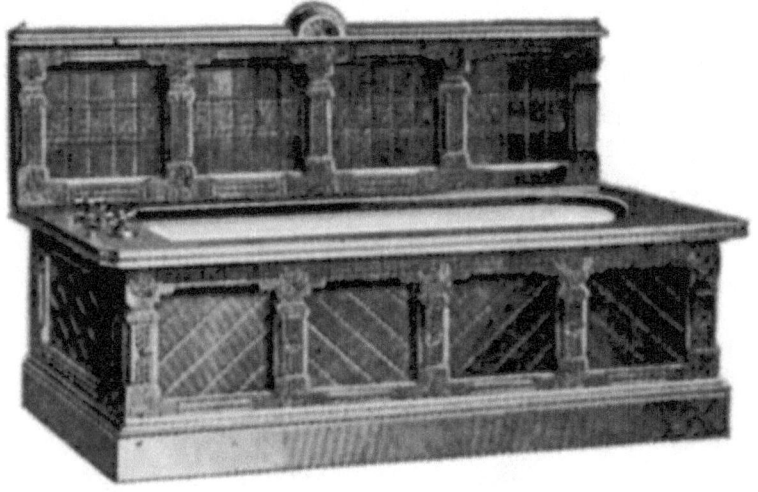

Fig. 6. — Encased bath tub.

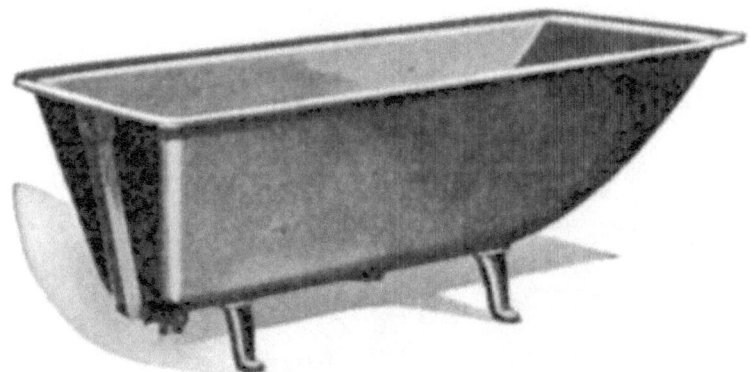

Fig. 7.—Steel tub on legs.

The Wash Bowl.—Succeeding the hand basin the first wash basins used in this country were made of marble or slate, with a round bowl of crockery. The bowl was 14 inches in diameter originally, but later was changed to an oval bowl. Like the bath tub these wash stands were [8] encased in wood, the encasing being used to support the marble top. Ornamental brackets were introduced and the wood encasement done away with.

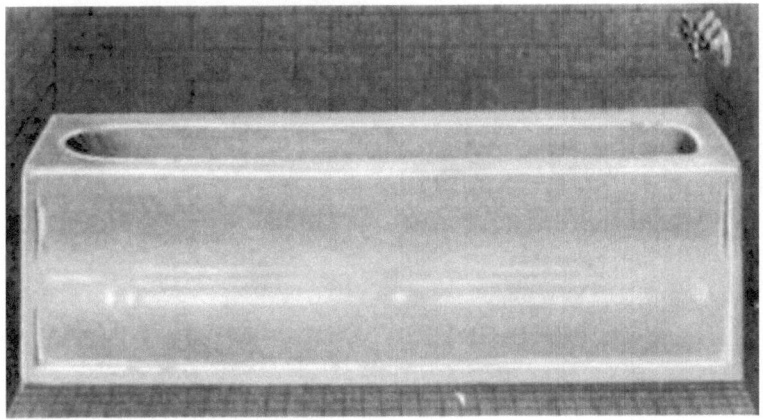

Fig. 8.—Modern built-in tub.

Fig. 9. — Encased wash bowl.

About 1902 the iron-enamelled lavatory appeared on the market and drove all other kinds from the market at once. The reason for this is clear. The marble stands were absorbent and were made with three parts, top, back, and bowl; the enamelled iron lavatory is made all in one piece of material non-absorbent. A study of the illustrations will show clearly how the lavatory has been improved. Strange to say, in all plumbing fixtures, and especially the [9] lavatory, as improvements were made to make them more sanitary a reduction has been made in the price of an individual fixture.

Fig. 10.

Fig. 11.—Bath room of early 80's. All fixtures are enclosed.

The development of the urinal, showers, wash trays, drinking fountains and other fixtures I will not attempt to cover. As the demand has been evident for fixtures of certain types, the plumber has been alert to anticipate and supply it. There is need, however, for improvement in [10] all our fixtures, especially that part which connects with the waste pipes, also the hanging, that is the arrangement or lack of arrangement for hanging fixtures to the wall. The waste and overflow of all fixtures need considerable change to make them sanitary. The opportunity is, therefore, before anyone who will apply himself to this development. Much money, thought, and time have been spent by the manufacturers of iron enamelled ware and by the potteries to gather suggestions made by the plumber in regard to fixtures, and then to perfect them. To these manufacturers is due the beautiful design, stability, and perfect sanitary material which make up our plumbing fixtures of today.

Fig. 12.

[11]

CHAPTER II

The Use and Care of the Soldering Iron. Fluxes. Making Different Soldering Joints

The Soldering Iron.—The soldering iron is one of the first tools a plumber has to master. This tool is sometimes called a "copper bit" as it is made of copper; and so throughout this book the words "soldering iron," "copper bit," "iron," and "bit" are used synonymously. There are several different-shaped irons in common use today, but an iron shaped like the one in Fig. 13 is the one for use in the following work. Take the iron as it is purchased, having a wooden handle and the copper exposed on pointed end. Before it can be used the point must be faced and tinned. To do this, proceed as follows:

1. *First*, heat the iron on the furnace.
2. *Second*, place in vise and file the four surfaces of the point.
3. *Third*, run a file over edges and point.
4. *Fourth*, heat the iron until it will melt solder.
5. *Fifth*, put 6 or 8 drops of solder and a piece of rosin the size of a chestnut on an ordinary red brick. (This rosin is called a flux.)
6. *Sixth*, take the hot iron and melt the solder and rosin on the brick.
7. *Seventh*, rub the four surfaces of the point of the iron on the brick keeping the point in the melted solder.

Fig. 13.—Copper.

The solder will soon stick to the copper surfaces and then [12] the iron is ready for use.

Another way to tin the iron that is in common use is to rub the point of a hot iron on a piece of sal-ammoniac, or dip the hot iron in reduced muriatic acid, then rub the stick of solder on the iron. The use of muriatic acid in tinning the iron is not recommended. In the first place, it is not always possible to carry it, and in the second place it eats holes in the surface of iron, which makes it necessary to file and smooth the surfaces again. The constant use of muriatic acid on the copper soon wears it away and makes it unfit for use. Rosin is easily carried and applied and is by far the best to use in regular work.

Points to Remember in the Care of the Soldering Iron.—

1. *First*, proper tinning is absolutely necessary for rapid and good work.
2. *Second*, do not allow the iron to heat red hot.
3. *Third*, keep the point of the iron properly shaped.
4. *Fourth*, use the same flux in tinning as is to be used in soldering.
5. *Fifth*, when filing iron, file as little as possible.
6. *Sixth*, keep in use two irons of the same size.

FLUX

A flux is used to clean the surfaces of joints and seams to be soldered, also to keep them from oxidizing and to help the metals to fuse.

The following list gives the names of various fluxes in common use, how they are applied, and on what material they are most commonly used:

Flux	How applied	Used on
Rosin	Sprinkled on	Lead, tin, and brass
Tallow	Melted	Lead and brass
Muriatic acid (reduced)	With swab	Copper, galvanized iron and brass
Muriatic acid (raw)	With swab	Dirty galvanized iron

Rosin.—Rosin is purchased by the pound and comes in [13] chunks. It is very brittle and powders easily. Plumbers generally take a piece of 1 1/4 N. P. brass tubing, solder a trap screw in one end and a cone-shaped piece of copper on the other. The point of the cone is left open. Rosin is put into this tube and is easily sprinkled on work when needed.

Tallow.—A plumber's *tallow candle* answers the purpose for tallow flux. Some plumbers carry a can for the tallow, making it cleaner to handle.

Muriatic Acid.—Muriatic acid or hydrochloric acid is used both raw and reduced. Raw acid is not diluted or reduced. Reduced acid is made as follows: Put some zinc chips in a lead receptacle and then pour in the muriatic acid. The acid will at once act on the zinc. The fumes should be allowed to escape into the outer air. When chemical action ceases, the liquid remaining is called reduced acid.

PLUMBERS' SOILS AND PASTE

It is necessary when soldering or wiping a joint to cover the parts of pipe adjoining the portion that is to be soldered or wiped so that the solder will not stick to it. There are a number of preparations for this. The one used by the best mechanics today is paste, made as follows:

- 8 teaspoons of flour.
- 1 teaspoon of salt.
- 1 teaspoon of sugar.
- Mix with water and boil down to a thick paste.

The advantages of paste as a soil are many:

1. *First*, it is made of materials easily obtained.
2. *Second*, solder will not stick to it.
3. *Third*, if pipe is thoroughly cleaned, the paste will not rub off easily.
4. *Fourth*, poor workmanship cannot be covered up.
5. *Fifth*, when the work is completed, a wet cloth will wipe [14] it off and leave the work clean.

Another soil used is *lampblack* and *glue*. A quantity of glue is melted and then lampblack is added. This needs to be heated and water added each time it is used. This soil is put on pipes with a short stubby brush. The work when completed with the silvery joint and jet black borders appears to the uninitiated very artistic and neat, but when the black soil is worn away the uneven edges of the joint appear, disclosing the reason for using a black soil that covers all defects. The mechanic of today who takes pride in his ability for good workmanship will not cover his work with black soil.

It can readily be seen that the use of lampblack soil encourages poor workmanship, while the use of paste forces, to a certain extent, good workmanship on the part of the mechanic.

Before soil or paste is applied, the pipe needs to be cleansed. Grease and dirt accumulate on the pipe. The methods employed to remove all foreign matter are simply to scrape the surface with fine sand or emery paper; sand and water will also answer for this purpose. This cleans the surface and allows the soil or paste to stick to the pipe.

MAKING DIFFERENT SOLDER JOINTS

The tools used in making the different solder joints as described and illustrated in this chapter are shown in Fig. 14.

Cup Joint.—The materials necessary for the work (Fig. 15): 12 inches of 1/2-inch AA lead pipe, paste, rosin, 1/2 and 1/2 solder.

If a gas furnace is not on the bench to heat the iron, then a gasoline furnace is necessary.

Each of the following operations must be done thoroughly to insure a perfect job:

1. *First*, with the **saw** cut off 12 inches of 1/2-inch AA lead [15] pipe from the coil. When cutting off a piece of lead pipe from a coil or reel, always straighten out 1 foot more than is needed. This leaves 1 foot of straight pipe always on the coil.

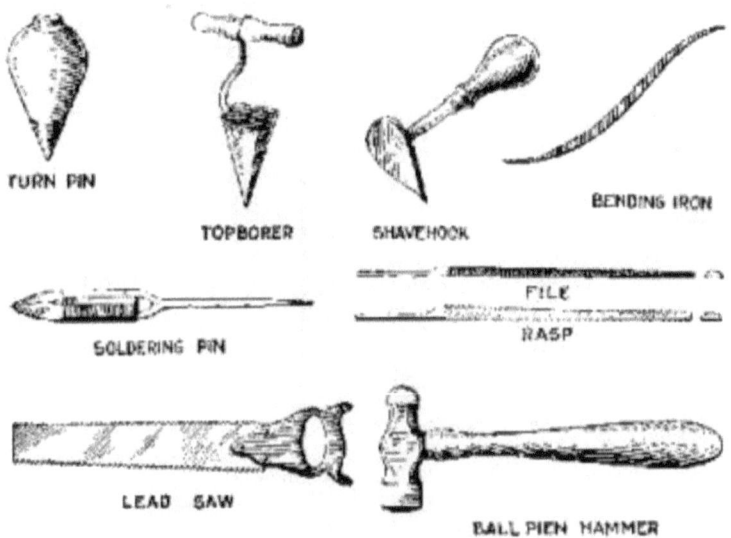

Fig. 14.—Tools used for making solder joints.

1. *Second*, with the flat side of the **rasp**, square the ends of the 12-inch piece of pipe. (A good way to do this is to hold the pipe at right angles with the edge of the bench, run the rasp across the end of the pipe, keeping the rasp *parallel* with the edge of the bench. Apply this to all work when necessary to square the ends of pipe.)

Third, cut the pipe with the saw, making two pieces each 6 inches in length.

Fourth, square the ends just cut.

Fifth, rasp the edges of one end as shown in the cut. Hold the work in such a way that the stroke of the rasp can be seen without moving the pipe.

Sixth, take the other 6-inch piece of pipe and with the [16] **turn pin** spread one end of it. The turn pin must be struck squarely in the center with the **hammer**, the point of the turn pin being kept in the center of the pipe. The pipe should be turned after each blow of the hammer. The pipe must not rest on the bench but should be held in the hand while using the turn pin. If the pipe bends, it can be straightened with **bending irons**. If the pipe is spread more on one side than the other, the turn pin should be hit on the opposite side so as to even the spread.

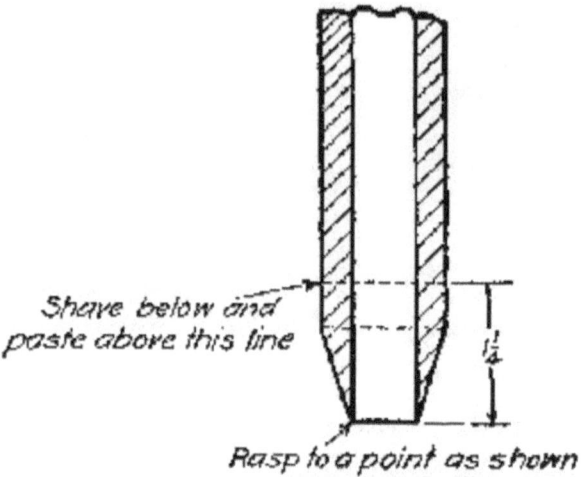

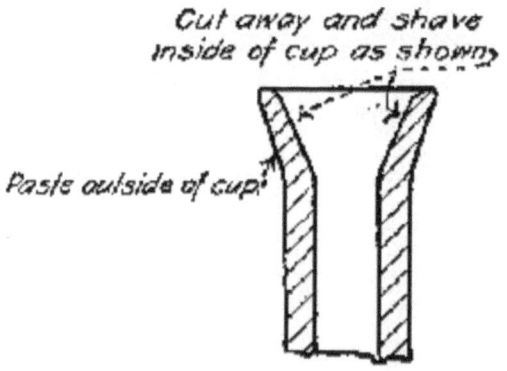

Fig. 15.

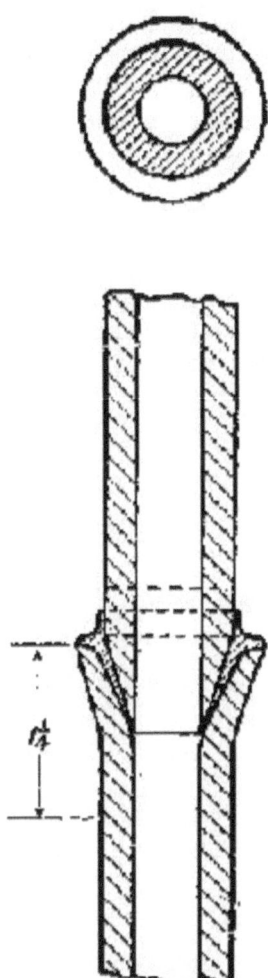

Fig. 16.—Cup joint.

1. *Seventh*, when the pipes are properly fitted, moisten the tips of the fingers with paste and rub the paste on parts of pipe marked "paste." Put the pipe aside to allow the paste to dry.

Eighth, put the soldering iron on to heat. [17]

Ninth, with the **shave hook** scrape off the paste and surface dirt as shown in the figure. The inside of the cup will look bright, but must be scraped.

Tenth, place the two pieces into position as shown in Fig. 16, sprinkle rosin on the joint, melt a few drops of solder on the joint and with the iron melt the solder on the joint, drawing the iron around the pipe keeping the solder melted around the iron all the time.

Eleventh, fill the joint with solder and continue to draw the hot iron around the joint until a smooth and bright surface is obtained. To master the correct use of the soldering iron in this work, considerable practice will be necessary.

Overcast Joints. — (Fig. 17.)

Note. — Each operation must be performed thoroughly.

1. *First*, saw off from a coil of 1½-inch D lead pipe a 10-inch piece of pipe.

Second, square the ends with the rasp, as previously explained.

Third, take a 1½-inch **drift plug** and drive through the pipe (Fig. 18).

Fourth, saw the pipe into two pieces of 5 inches each.

Fifth, square the ends of the pipe with the rasp.

Sixth, rasp off the outside edge of one end of the pipe as shown.

Seventh, rasp off the inside edge of one end of the pipe.

Eighth, finish rasped surfaces with a file. Both surfaces should have the same angle.

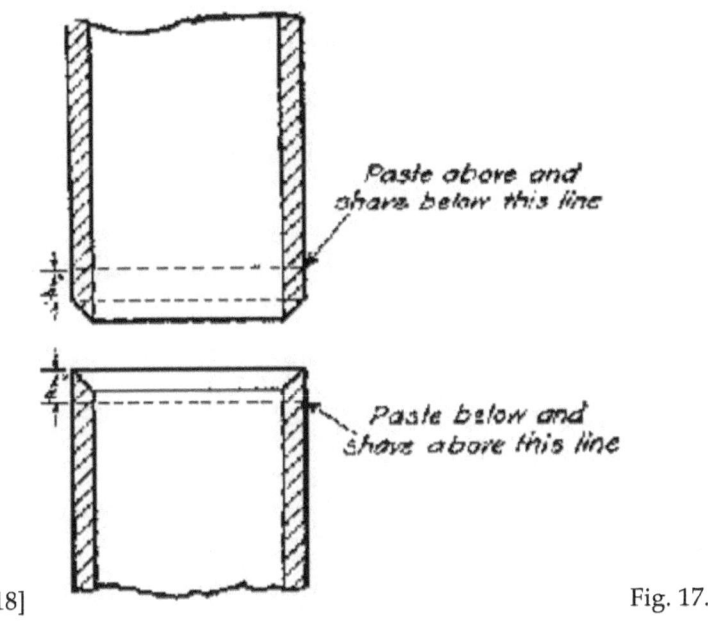

Fig. 17.

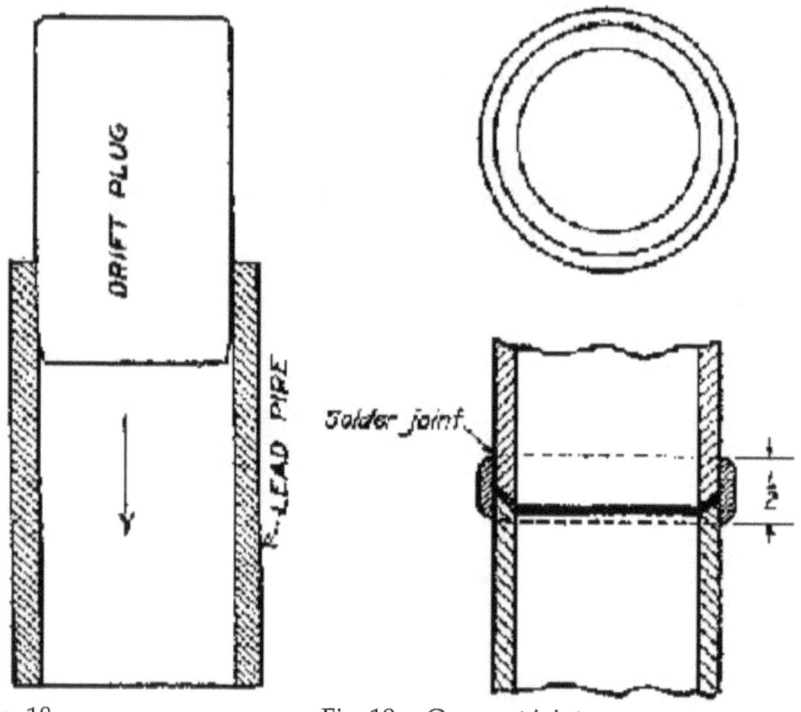

Fig. 18. Fig. 19.—Overcast joint.

1. *Ninth*, with a shave hook scrape the outside surface of [19] each pipe for about 1 inch from the end.

Tenth, put the soldering iron on to heat.

Eleventh, paste paper on the joint as shown in the cut.

Twelfth, fit the pieces together and lay on the bench. Drop some melted solder on the joint and with the hot iron proceed to flow the solder around the joint by turning the pipe. Use plenty of flux (rosin). The pipes must be tacked in three or four places at first or they will have a tendency to spread.

Thirteenth, to finish the joint, lift the iron straight up.

This joint when finished will have a bright smooth finish. The two foregoing joints need considerable practice and should be perfectly mastered before going on to the next job.

SEAMS

A description of the making of wiped seams for lead-lined tanks will not be attempted as very few are made now. The plumber, however, is often called upon to make a seam joining two pieces of sheet lead. The beginner will do well to go over the following exercise carefully and practice it thoroughly.

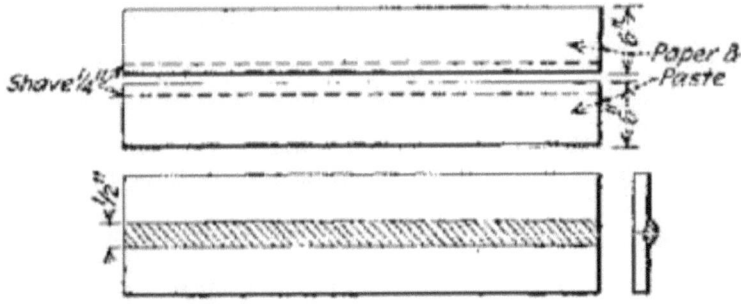

Fig. 20.—Flat seam.

Materials.—Two pieces of 8-pound sheet lead, 6 by 10 inches each; one bar of 1/2 and 1/2 solder; paste, paper, and rosin.

Tools.—Rasp, shave hook, and soldering iron. [20]

The 10-inch side of each piece is rasped and fitted together. The edges are cleaned and paper is pasted on leaving 1/4 inch for solder. Paste without the paper can be put on. This will make a joint 1/2 inch wide.

Apply the rosin to the joint, then with the heated iron and some solder tack the seam on the top, then on the bottom and middle. This will prevent the seam from spreading when the lead is heated. Solder and rosin can now be put on the full length of the joint. With a hot iron proceed to float the solder down the seam. The soldering iron must not rest at full length on the pieces of lead or it will melt the lead and render the work useless. The solder will flow and form a clean neat seam, if the iron is at the right heat and the right amount of solder is put on. If the iron is too hot, the solder will flow instantly when the iron is laid on it and the solder will disappear as it runs through the seam. If the iron is too cold the solder will not melt enough to flow. Too much solder on the seam will cause it to overflow, that is, the solder will spread beyond the papered edges. After a little practice this surplus solder can be drawn in on the seam with the iron and carried along the seam to some point that has not enough solder. When the seam is completed the edges should be perfectly straight and even. The iron is carried along the seam with one stroke which makes the seam appear smooth and bright.

[21]

CHAPTER III

Mixtures of Solders for Soldering Iron and Wiping. Care of Solders. Melting Points of Metals and Alloys

The importance of good solder, that is, solder correctly mixed and thoroughly cleaned, should not be overlooked. Work is more quickly and neatly done and the job presents a more finished appearance when solder that is correctly made is used.

The solder used in the following work with the soldering iron is called 1/2 and 1/2. This means 1/2 (50 per cent.) lead and 1/2 (50 per cent.) tin.

In the mixture of solder, only pure metals should be used. The lead should be melted first and all the dross cleaned off. The tin should then be added and mixed.

The solder to be used in wiping the joints in the following chapter is a mixture of 37 per cent. tin and 63 per cent. lead. This is called wiping solder.

The following table gives the melting points, etc.:

Metal	Melting point	Mixture
Sulphur............	228	Pure
Tin....................	446	Pure
Lead.................	626	Pure
Zinc..................	680	Pure
Fine solder........	400	50 per cent. tin, 50 per cent. lead (wt.)
Wiping solder....	370	37 per cent. tin, 63 per cent. lead (wt.)

To recognize fine solder, run off a bar into a mold and let [22] it cool. If there is a frosted streak in the center, the metal has not enough tin. The surface should be bright. To recognize wiping solder, pour some on a brick. When this is cool, the top should be frosty and the under side should have four or five bright spots. The amount poured on the brick should be about the size of a half dollar. If poured on iron, the metal will cool too quickly and show bright all over the under side.

To Make 1/2 and 1/2 Solder or Plumber's Fine Solder.—The possibility of getting pure clean metals to mix solder is very remote. Old pieces of lead pipe, lead trap, old block tin pipe are used to make solder when pure metals are not at hand.

> 1. *First*, in a cast-iron pot melt the lead to about 800°, or a dull red.
>
> *Second*, clean off the dross.
>
> *Third*, add (to a 15-pound pot) 1/2 pound of sulphur in three applications. Each time mix the sulphur thoroughly with the metal with a long stick.
>
> *Fourth*, add tin before the last application of sulphur. Mix thoroughly.
>
> *Fifth*, pour off two bars and look for the frosty streak in the center. Add a little more tin, if necessary.

To Mix Wiping Solder.—

> 1. *First*, proceed as described in 1/2 and 1/2, melting the metals and *burning* out with sulphur, adding the percentage of tin according to the preceding table. Then test the solder for bright spots on the under side.
>
> *Second*, keep the metal thoroughly mixed when burning and keep all dross cleaned off the surface.

The working heat of wiping solder is 500°F. Sulphur is used to collect all zinc and dross. The sulphur should come in contact with all parts of the metal. This is why the metal should be stirred when the sulphur is put in.

A few good points in the economical care of solder are [23] listed below.

Care of 1/2 and 1/2 Solder.—

1. *First*, do not drop melted solder on the floor or dirty bench.

Second, use all small ends by melting on a new bar.

Third, put clean paper under work and use droppings.

Fourth, have the mold free from dirt when pouring.

Care of Wiping Solder.—

1. *First*, do not heat red hot.

Second, do not file brass where the filings will get into the solder.

Third, do not allow lead chips to get into the solder.

Fourth, clean the solder occasionally.

Fifth, learn to distinguish solder from lead by its hardness.

Sixth, have different-shaped pot for lead and solder.

Seventh, do not *tin* brass by dipping into solder.

Eighth, do not put cold or wet ladle into hot solder.

A pot holding about 15 pounds of solder is the size commonly in use.

[24]

CHAPTER IV

Making and Care of Wiping Cloths

A good wiping cloth is essential for wiping joints. The exact size and the flexibility of the cloth depend a great deal upon the mechanic who handles the cloth. Some mechanics like a stiff cloth, but the writer has always used a flexible cloth. The sizes, shape, and methods of folding and breaking in as shown in Fig. 21 below have proved successful. Cloths made of whalebone ticking are inexpensive and make the best for ordinary use.

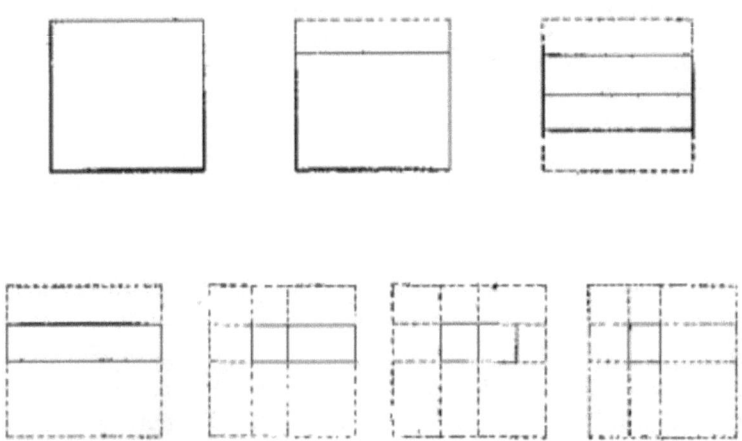

Fig. 21.—Folding a wiping cloth.

Size of cloth open		Size of cloth folded
14 1/2 by 14 1/2 inches	equals	3 1/4 by 3 1/4 inches
13 1/2 by 13 1/2 inches	equals	3 by 3 inches
8 1/2 by 12 1/2 inches	equals	2 by 3 inches

For the joint-wiping jobs to follow, the above sizes are [25] the best. The largest size, 14 1/2 by 14 1/2 inches is used for *catch cloth*. The 13 1/2 by 13 1/2 inches is the *wiping cloth*. The 8 1/2 by 12 1/2 inches is the *branch cloth*.

Proceed as follows to cut and complete a cloth:

1. *First*, lay the ticking on the flat bench and square the sides 14 1/2 by 14 1/2 inches.
2. *Second*, the ticking should be cut off with shears and not torn or cut with a knife.
3. *Third*, fold as shown in the cut.

Each fold should be moistened with a little water and pressed with a hot iron. The cloth should not be pulled or stretched, but should be kept as square as possible.

The first and second folds require a little care; the corners when folded to the center should be kept in a little, thus making the outside edge slightly rounded. If this is done, the corners will not stick out when the cloth is finished. After the cloth is carefully folded, pressed, and dried, take a needle and thread and sew the open corners about 1/2 inch in from the edge of the cloth. By carefully studying the cut, one can readily see each operation and, by following directions, make a perfect cloth.

When the cloth is done, an amount of oil sufficient to soak through about three layers of cloth should be applied and then rubbed on a smooth surface. The oil should be rubbed in well about the edges. It will not be necessary to apply anything else to the cloth to prepare it for wiping. Paste, soil, chalk, etc., are not needed and do not benefit the cloth. When using oil on the cloth, it must not be used too freely, that is, the cloth must not be soaked in oil, as oil is a rapid conductor of heat and the cloth would soon become too hot to handle.

Care of Wiping Cloths. — The ticking will burn if allowed to become too hot. If hot solder is poured directly on the cloth, it will soon burn and be destroyed.

Keep the surface on both sides of the cloth well oiled. [26]

Use both sides of the cloth.

Use both wiping edges of the cloth.

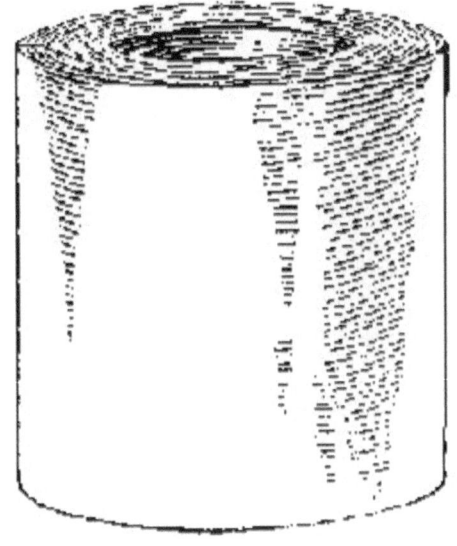

Fig. 22. — Wiping cloth folded has 16 thicknesses of ticking.

When the cloth is not in use, it should not be thrown in with the other tools and allowed to curl up into all sorts of shapes, but should be kept in some flat place. A good way to keep the cloths is to have two pieces of wood between which the cloths may be kept and held there by means of a strap. The length of time which a wiping cloth can be used depends a great deal upon its making and upon the care which is given it.

[27]

CHAPTER V

Preparing and Wiping Joints

When the writer first started to carry the tools for a plumber and to prepare joints for wiping, the remark was often heard that joint wiping would soon be a thing of the past. I have heard this many times since from many different sources. Personally, I fail to see the passing of the wiped joint. More lead pipe is being made today than ever before, which goes to show that lead pipe is being used and the only successful way of joining is with the wiped joint. Some plumbers' helpers of today seem to think that joint wiping is of no account. To a certain extent, I can sympathize with them. Most of these boys are learning a trade in large cities and working for concerns that do nothing but a large contracting business. This large work is carried on differently from the small work. Wrought-iron or steel pipes are used to a great extent in this work and a very small amount of lead is used. Sometimes the job will be completed without the use of lead. The boy who works continually on this kind of work soon comes to think that lead pipes are no longer in use. The writer has found that a boy who has learned to do nothing but screw-pipe work is absolutely lost and cannot perform the duties of a plumber, other than screw-pipe work. It must be borne in mind that lead pipe and cast-iron pipe work are being used today in all parts of the country and in some parts more than in others. Therefore, the boy must grasp all branches of the trade that he has chosen to follow and not be a one-sided man. Joint wiping belongs to the plumber alone. The plumbing trade differs from all other trades in that it has [28] joint wiping for its distinctive feature.

A few attempts at joint wiping will convince the beginner that it is not the easiest thing in the world to learn. Let me caution the beginner not to get discouraged. He must have patience and a firm resolve to master the art of joint wiping and not let it master him and keep him back.

So, as we now start on exercises of joint wiping, let the beginner constantly keep in mind that all boys must become perfectly skilled in the art of joint wiping before they can be considered plumbers. Keep in mind also that the examination that one must take to get a plumber's license contains an actual exercise in joint wiping. The one word of advice is not to get discouraged. Continued practice is the only way to success.

The soldering iron is, or should be, conquered by this time. As joint wiping is the next exercise, I shall go over a few general points that experience has taught me and cannot fail to be of assistance to the beginner if they are heeded. In fact, to become proficient, the beginner should remember all the points suggested under this heading. It is necessary in wiping to have good solder. In the chapter on solder, I have given the correct mixtures and how to recognize the proper mixtures. The place where wiping is to be done should be considered. No draught should be allowed to blow across the work as it tends to chill the solder and pipe. Proper support for the work should be procured. If gasoline is to be used for fuel to heat the solder, make sure that the tank is full before starting, otherwise the fire may go out just when the heat is needed most and the solder in the pot has become too cool to wipe with. Have a catch pan and keep all the solder droppings to put back into the pot, otherwise the solder will pile up and the fingers are likely to be pushed into the pile and badly burned. Hold the ladle about 2 inches above the work, the catch cloth about 1 inches below. Do not drop the solder in the same [29] place. Keep moving the ladle. Do not pour the solder on the pipe in a steady stream, but drop it on. It is not a large amount of solder that is wanted on the joint at first, it is heat that is needed. This can be secured better by dropping the solder on than by pouring a large quantity on the pipe. The edges of the joint cool very quickly; therefore heat the edges well and keep them covered with molten solder until the joint is ready to wipe. When preparing joints for wiping, always do the work thoroughly and fit the pieces together tightly so that no solder can get through.

Points to Remember.—

1. *First*, good solder.
2. *Second*, place of wiping.
3. *Third*, support.
4. *Fourth*, full tank of gasoline.
5. *Fifth*, drip pan.
6. *Sixth*, ladle 2 inches above the work.

7. *Seventh*, cloth 1 inches below the work.
8. *Eighth*, move the ladle continually.
9. *Ninth, drop* the solder.
10. *Tenth, heat,* not solder wanted at first.
11. *Eleventh*, heat the edges.
12. *Twelfth*, careful preparation.
13. *Thirteenth*, clean grease from the pipe.
14. *Fourteenth*, cut clean straight edges on paper.

HALF-INCH ROUND WIPED JOINT

Preparation.—Take 12 inches of 1/2-inch strong lead pipe and square off the ends with a rasp. Take the shave hook and scrape the center of the pipe perfectly bright; a space 3 inches each side of the center is correct. The size of the joint when completed should be 2 1/2 inches long. If we should undertake to wipe the joint with the pipe in the [30] present condition, the solder would adhere to all the pipe that was shaved bright. Therefore, we take a piece of paper sufficient to encircle the pipe twice and after putting paste on one side of the paper wrap it around the pipe so that the edge that is cut straight and even is 1 1/4 inches from the center of the pipe. Another piece of paper is pasted on the other side of the center leaving a clean, bright space of 2 1/2 inches. All the pipe should be covered with paper except the 2 1/2 inches in the center.

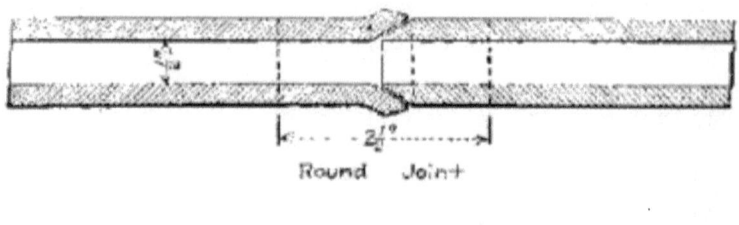

Round Joint

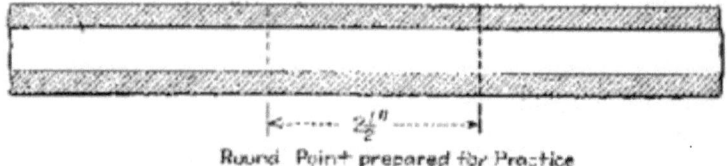

Round Point prepared for Practice

Fig. 23.

To Put the Pipe in Position for Wiping.—The most practical way is to take two common red bricks with the 2 by 8 face down and place them 9 inches apart. Lay the pipe on the bricks and place a weight on each end. The solder will drop on to the bench, so it is best to place a piece of paper or a pan of black iron under the pipe to catch the solder that drops. The pan or paper can then be taken up and the solder put back into the pot without waste. A cast-iron pot holding 15 pounds of solder is then placed on the furnace. When the solder has melted and has reached 500° it is ready for use. This can best be determined by putting a piece of paper in the solder. If the paper scorches, the solder is at the right heat; if the paper catches fire, it is too hot.

Now take a 3-inch ladle and heat it over the fire and then dip it into the solder and skim off any dross that may have collected.

Wiping.—With the ladle full of solder in the right hand and the large cloth or the catch cloth in the left hand, begin to drop the solder on the joint. The cloth should catch all the solder as it falls off the pipe. If hot solder is held against the bottom of the pipe, it is heated to the proper heat. Always begin to drop the solder on the paper edges, then drop the solder on the joint itself. Bear in mind that the solder should not be poured on, but dropped on slowly. After the first few drops do not drop the solder directly on to the lead pipe but on to the solder previously put on the pipe. This will save the pipe from burning through. The pipe must be the same heat as the solder before the proper heat is obtained for good wiping. The beginner should practice dropping the solder on the joint, catching the solder and working it around the pipe. By doing this, one becomes familiar with the feeling of hot solder, which is the secret of successful wiping. When the solder works easily around the pipe, drop the ladle and take the smaller wiping cloth in the right hand and with both cloths draw all the solder on top of the pipe. With fingers on the corners of both cloths, clean off the left-hand edge and with the right hand draw the surplus solder across to the right-hand edge. Next, clean the right-hand edge of the joint pushing the surplus solder onto the cloth in the right hand. Work this solder on to the bottom of the joint. Now discard the catch cloth. Holding the wiping cloth with the index [32] fingers on lower opposite corners, shape the under and front side of the joint. With the middle fingers on opposite lower corners of the cloth shape the back and top. Keep the index and middle fingers on the edge of the cloth and the edge of the cloth on the edge of the joint. This position together with the size and shape of the cloth will give the joint the desired form and appearance. Particular attention is called to the position of the fingers as shown in the figure.

The last wipe should be a quick stroke coming off of joint on a tangent. If the solder is at right heat, the cloth will not leave a noticeable mark. If, however, the solder is too cold, a ragged edge will result. Sometimes a cross wipe is made for the last stroke and a good finish obtained.

Points to Remember.—

1. *First*, width of the joint, 2½ inches.
2. *Second*, allow no soil or paste to get on the joint.
3. *Third*, a 3-inch ladle should be used.
4. *Fourth*, 500° is the working heat of solder.
5. *Fifth*, paper test for solder heat.

6. *Sixth*, position of wiping cloths.
7. *Seventh*, do not drop solder on the lead pipe.
8. *Eighth*, hold the ladle 2 inches above the pipe.
9. *Ninth*, wipe the edges of the joint first.
10. *Tenth*, wipe and shape the joint quickly.

The above procedure of wiping will be found to work out very easily if followed closely. Do not pour the hot solder onto the cloth as the cloth will burn through and soon be useless. A little more oil should be put on the cloth after using it for awhile. The cloth should be turned around and the opposite side also used. The cloth will last considerably longer if sides are changed frequently. The solder should not accumulate on the pan, but should be continually put back into the pot. The "metal," as solder is sometimes called, should never be allowed to become [33] red hot.

The above method of preparing pipe is suggested for beginners only and will be found to be a great help to them. In actual practice the joint must be prepared differently. The method used in trade is as follows:

The joint is used to join two pieces of lead pipe. Take two pieces and rasp the four ends square. With the tap borer clean out the end of one pipe a trifle, then with the turn pin enlarge this end just a little as shown in the figure. Then rasp the edge off about 1/8 inch as shown. Take the other piece of pipe and rasp one end as was done in the cup joint, making it fit into the first piece. Then place the two ends together and with the bending iron beat the pipe, making the joint as tight as possible.

ROUND JOINT — 45° TO RIGHT

The next position in which the beginner is to wipe a joint is on an angle of 45° to the right.

Preparation.—To prepare this joint, proceed as in the horizontal round joint. I will enumerate a few of these points. A piece 12 inches long of 1/2-inch pipe is cut off and the ends squared. A strip in the center, 6 inches long, is shaved clean. Paper and paste are put over the pipe except 2 1/2 inches in the center. Grease can be put on the pipe in between the pieces of paper and will keep the lead from oxidizing.

Placing Pipe in Position. — There is no need of an elaborate system of holding the pipe in position. Take a red brick and place the 4 by 8 face down. This will do for the bottom pipe. For the top of pipe to rest on, place two bricks one above the other; this will give the correct position. Place the pipe on the brick and with a ladle full of half molten solder pour a clamp of solder over the end of the pipe. This will hold the pipe firm for wiping. Place a [34] catch pan under the joint for solder to fall in.

Wiping. — The method of wiping this joint is practically the same as wiping the horizontal joint. The catch cloth should be held parallel with the bench tilting a little from front toward the back. The ladle is held the same and solder is dropped on as before. The ladle should be continually moving while dropping solder, not allowing the solder to drop twice in the same place. When the solder has been worked around the pipe and is at working heat, the solder is drawn up with both cloths and the top edge wiped first, then the bottom edge; the surplus solder is put on the underside of the joint, and then with three or four wipes the joint is made symmetrical and finished.

Things to Remember. —

1. *First*, prepare like the horizontal joint.
2. *Second*, use brick to place in position.
3. *Third*, hold tools as in horizontal joint.
4. *Fourth*, top edge cools first, therefore, wipe it first.
5. *Fifth*, hold the wiping cloth at an angle of 45° when wiping, with fingers placed as noted in previous joint.
6. *Sixth*, make solder clamp for holding the pipe.

ROUND JOINT 45° — LEFT

When the preceding joint is well mastered and a number of good joints have been wiped, turn the pipe on an angle of 45° to the left.

Preparation. — The preparation for this joint is exactly the same as for the horizontal joint. The beginner should turn back and read carefully concerning the perfection of the joint. Bear in mind that the pipe must be correctly prepared or a good joint cannot be made. The edge of the paper must be cut not torn.

Placing Pipe in Position.—This pipe can be placed in position the same as the preceding one. If heavy weights are placed on the ends of the pipe, a bad habit may be [35] formed by the one learning to wipe. That is, the habit of pressing hard on the joint when wiping. In the preceding joint, if the beginner presses too hard, the pipe will fall off the bricks.

Wiping.—Proceed as described for previous joints. The top edge must be favored a little. The hot solder will run down to the bottom edge; therefore less solder should be dropped on it than on the top edge. When the solder is at the proper heat for wiping it requires only a light touch to wipe the joint. If it appears necessary to press hard on the joint to wipe off surplus solder, it shows that the solder is not at the correct wiping heat.

ROUND JOINT—VERTICAL

Preparation.—This joint can be prepared exactly like the preceding one. In fact, the same piece of pipe can be used throughout. When preparing this joint the end that is to be on the bottom should be well covered with paper.

Placing in Position.—The best way to hold this joint in position for wiping is to stand the pipe upright on one end with the pan underneath. A piece of furring strip should be run from the top of the pipe to the wall. Secure the strip to the wall and drive a nail through the strip into the bore of the pipe. Place a weight on top of the strip and the pipe is ready.

Wiping.—The procedure of wiping this joint is entirely different from that in the other positions. The solder is thrown onto the joint from the ladle. The catch cloth is held up to the pipe and as much solder as possible is held on to the pipe. Move the ladle around the joint, throwing a little solder on as the ladle is moved. Notice now that all the solder runs to the bottom edge, leaving the top edge cold. The solder that accumulates on the bottom edge should be drawn up to the top edge with the cloth. Then [36] splash more solder on to the top edge and as the solder runs down the pipe catch it with the cloth and draw it up again. The solder can be worked around and up and down the joint, but always keep the top edge covered with hot solder. The solder is likely to drop off the joint entirely unless watched closely. When the correct heat is obtained, drop the ladle. Take the wiping cloth in the right hand and with the fingers spread, clean off the top edge quickly, then shape the joint with the one cloth. With a little practice you will gain this knack. The joint can then be wiped. The left hand can steady the pipe. Spread the index finger and third finger to opposite sides of the cloth and wipe around the joint.

[37]

CHAPTER VI

Preparing and Wiping Joints (*Continued*)

TWO-INCH BRASS FERRULE

Materials.—The beginner should continue wiping the vertical round joint until he is able to obtain a symmetrical bulb. A joint should be wiped in each of the foregoing positions for exhibition purposes, so that the beginner can have before him the best work and strive to make the next joint better. This next joint, the 2-inch brass ferrule, is wiped in an upright position. The materials necessary are the 2-inch brass ferrule, 6 inches of 2-inch light lead pipe, paste and paper, 1/2 and 1/2 solder, rosin, wiping solder, catch pan, and supports.

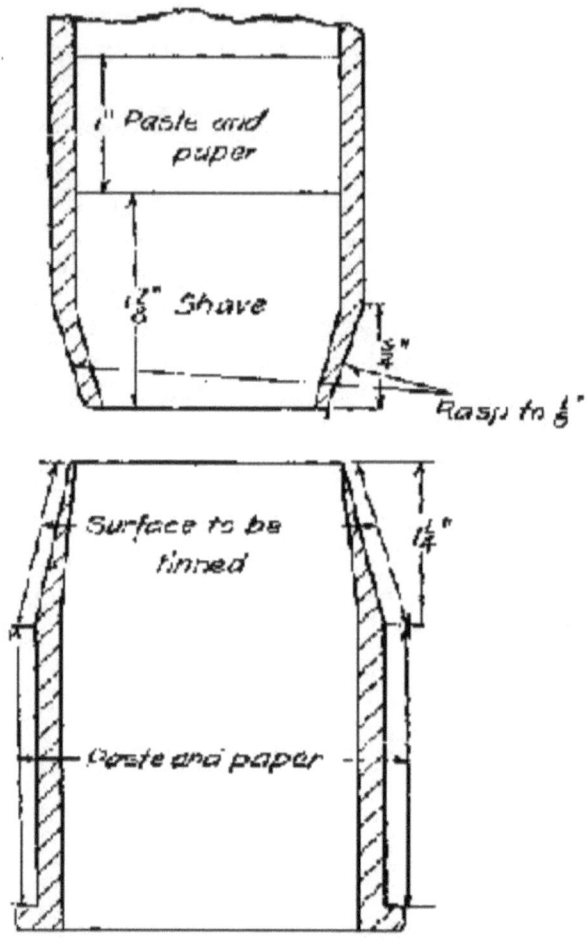

Fig. 24.

Tools Required.—The tools necessary for this work are as follows: the saw, rasp, drift plug, dresser, file, soldering iron, bending irons, wiping cloths, shave hook, and ladle.

Preparation.—The lead pipe must be fitted into the brass ferrule. The brass ferrule has to be tinned first. To do [38] this, proceed as follows: file the ferrule for about 2 inches on the tapered end. Do not file too deep, but just enough to expose the pure bright metal. Now measure from the small end 1 1/4 inches down toward the beaded end. From this point to the bead, cover the brass with paste and paper. No paste must get on the 1 1/4-in. filed end. This end should not be touched with the fingers. If paste gets on it, the process of filing must be done over again as the solder will not stick where there is paste. If the brass ferrule is filed while the paper is on the brass, the filing will destroy the straight edge of the paper and an even joint cannot be made. It would therefore be necessary to re-paper the brass. Take some powdered rosin and cover the filed end of the ferrule with molten solder using the rosin as a flux. Do not dip the end of the ferrule into the hot wiping solder to tin it or pour wiping solder on the brass ferrule. This method of tinning the ferrule will spoil the wiping solder. Always use the soldering iron to tin the ferrule as explained above. A little practice will develop the use of the iron in the hands of the beginner so that this tinning process will be done very rapidly. The iron should be put on to heat when the paper is being pasted on the brass; the iron will then be ready for use when needed.

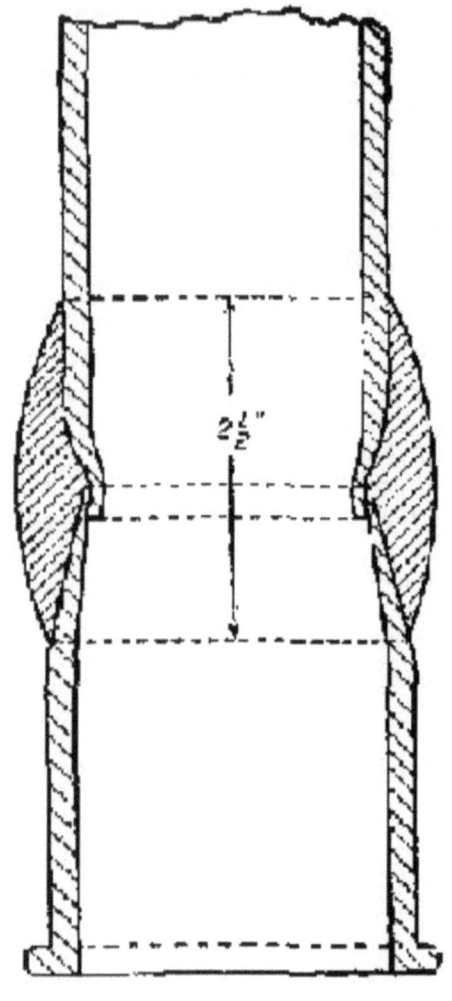

Fig. 25.—Two-inch brass ferrule.

Preparing the Lead.—The ends of the lead pipe must be squared with the rasp. All kinks and dents are taken out by using the drift plug and driving it through the pipe. Take a piece of smooth pine stick and start to beat in the end of the lead pipe to fit the brass ferrule. The pipe should be beaten in starting about 3/4 inches from the end. It should be beaten in very slowly until it fits the ferrule. The pipe is held in the hand all the time and considerable time should be spent on this as it is the first time the beating in of lead pipe has been called for. The knack of doing this comes only by slow and continued practice. The lead must be "humored" into shape and not "driven" into shape. [39] The end of the pipe is tapered still more by rasping off the end. About 3/4 inch should extend into the brass ferrule. With the bending irons, the lead extending into the brass ferrule is beaten against the inside wall of the ferrule. A good way to do this is to wedge the lead pipe in as much as possible at first, then lay the work flat on the bench, in which position it is more easily worked. The sketch should be thoroughly studied and each notation be perfectly understood, before proceeding with the work. Now that the lead pipe is perfectly fitted into place, it is prepared for wiping. The joint overall will be 2 1/2 inches. As we have already allowed 1 1/4 inches on the brass ferrule for the joint, the lead will have to be cleaned that much more. With the shave hook, shave the end of the pipe that has been fitted into the brass ferrule. A space about 4 inches should be cleaned. This will give a cleaned surface free from dirt and grease for the paste and paper to adhere to. Next paste the paper in place. The lead pipe can be entirely covered, or 3 or 4 inches only, above the 1 1/4 inches allowed for the joint. The space between the paper on the brass and the paper on the lead should now be 2 1/2 inches. The paste and paper should now be allowed to dry.

Supporting the Pipe.—This joint is wiped with the ferrule down on the bench. A flat pan is laid on the bench and the ferrule stood upon it. A weight on top of the lead pipe is all that is necessary. If this does not make the pipe rigid enough for the beginner, then a support similar to the round vertical joint support can be used. The beginner is advised, however, to practice the wiping of this joint with only the weight to hold it in position. The beginner will then be required to wipe the joint while the solder is hot, when it does not require a heavy pressure against the solder to wipe it in shape. These wiped joints should be supported in place near the furnace that heats the solder so that the solder will be handy for [40] wiping.

Wiping.—Wiping this joint brings in some of the methods of the round vertical joint. If that joint was thoroughly mastered, this joint will be wiped considerably more easily. The ladle is held in the right hand and the solder splashed on the joint. The catch cloth is held in the left hand and some of the solder is caught and brought up on the top edge. The top edge cools quickly as all the hot solder runs down to the bottom edge and into the pan. As the solder accumulates on the bottom edge, it is drawn up on the top edge, and in this manner the top edge is kept hot. When the solder can be worked freely around the pipe and the edges are hot, the joint is ready to wipe. The ladle is laid down and the wiping cloth is taken in the right hand and the top edge of the joint cleaned on one side. Then the wiping cloth is changed to the left hand and the other side of the top edge is cleaned. Holding the cloth in one hand with the index and the third fingers spread to the outside corners of the cloth, the cloth is passed around the joint quickly. To get an even and symmetrical joint, it is necessary to make two or three passes around the joint holding the cloth first in the right and then in the left hand. The free hand is used to steady the work. This joint should be wiped very slim to allow room for the caulking irons to pass by it and get into the hub of the pipe. Constant wiping on the brass ferrule will [41] result in the tinning on the brass ferrule coming off. The ferrule will look black when this happens and will thus be recognized. The wiping should then be stopped and the ferrule filed and tinned in the same manner as it was done at first.

Points to Remember.—

1. *First*, material—6 inches of 2-inch light lead pipe and one 2-inch brass ferrule.
2. *Second*, tin ferrule, using soldering iron.
3. *Third*, use a soft pine stick for a dresser.
4. *Fourth*, fit the lead into the ferrule.
5. *Fifth*, clean and paper the lead.
6. *Sixth*, secure the pipe into position.
7. *Seventh*, using the catch cloth and ladle, splash solder on the joint.
8. *Eighth*, keep the top edge covered with solder.
9. *Ninth*, wipe the top edge first.
10. *Tenth*, shape and finish wiping with a few strokes.
11. *Eleventh*, tools used.

12. *Twelfth*, wipe a slim joint.
13. *Thirteenth*, steady the work with the free hand.
14. *Fourteenth*, re-tin the ferrule, if necessary.

FOUR-INCH BRASS FERRULE

The 4-inch brass ferrule joint is the same as the 2-inch, except for size. The materials needed for this joint are 6 inches of 4-inch, 8-pound lead pipe, and one 4-inch brass ferrule, one *full* pot of solder, some paste and paper, rosin, and 1/2 and 1/2 solder.

Tools Necessary.—The tools required for this joint are as follows: saw, rasp, file, ladle, soldering iron, dresser, bending irons, shave hook, and wiping cloths.

Preparation.—*Lead Pipe.*—With the saw cut off 6 inches of 4-inch lead pipe. This pipe comes in lengths and should [42] be for this work about 8 pounds to the foot in weight. The pipe may be dented badly, but these dents can be taken out as follows: Take a piece of 2-inch iron pipe and put it in a vise. The lead pipe can be slipped over this iron pipe and any dents taken out easily by beating with the dresser. One end of the lead pipe is beaten with the dresser until it fits into the ferrule. The end is then rasped a little. Then, after the brass ferrule has been tinned, the pipe is fitted into it and beaten out against the inside wall of the brass ferrule and a tight joint is made. The lead is next cleaned with the shave hook and paper is pasted on as explained under the 2-inch brass ferrule, the description of which should now be read over.

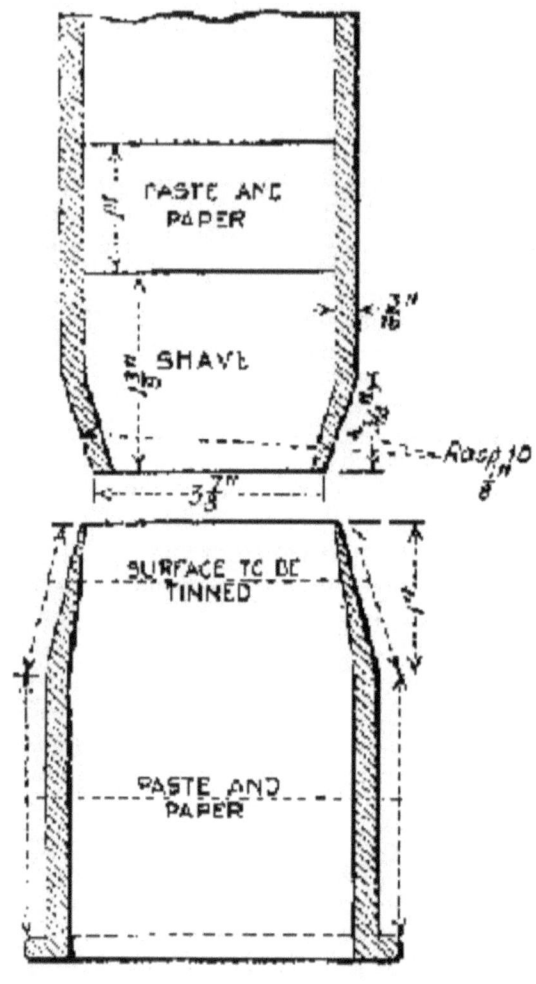

Fig. 26.—Four-inch brass ferrule.

Brass Ferrule.—The first thing to do with the brass ferrule is to file the end that is to be wiped. When the brass ferrule is filed, it should be done away from any part of the room where the filings are likely to get into the solder. After the filing has been done, paper is pasted on all of it except the part that is to be tinned and no paste must get on to this part of the ferrule. If any paste does get on to it, the filing will have to be done over again. When using paste and paper, neatness must be cultivated, or paste will be spread over parts of the pipe that are supposed not to have any paste on them. Next, take the soldering iron and heat it. Take some rosin and put it on the exposed part of the ferrule. With the hot soldering iron [43] proceed to tin the brass ferrule, as explained before, with 1/2 and 1/2 solder, using rosin as a flux. Now the lead pipe that has previously been prepared is fitted into the ferrule.

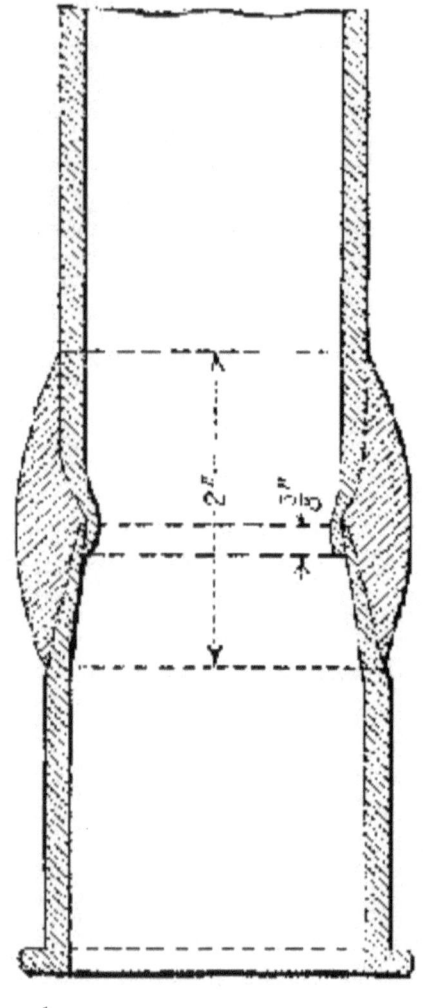

Fig. 27.—Four-inch brass ferrule.

Supporting.—Set the brass ferrule on a catch pan. The lead pipe is upright. A weight placed on top of the lead pipe will steady the pipe for wiping. When the joint is wiped the free hand can hold the pipe if the weight is not sufficient to support it.

Wiping.—Splash the solder on the joint from the ladle, in the same manner as was employed in the two preceding joints. To get the proper heat on the 4-inch joint a little more speed is necessary, also the constant working of the solder around the pipe. The ladle is constantly moved around the pipe so that all parts of the pipe will be evenly heated and come into contact with the hot solder direct from the ladle. When the solder works freely around the pipe and the top edge is hot, the joint is shaped by holding the wiping cloth in the right hand, with the index and the middle fingers spread to the opposite corners of the cloth. The fingers are placed one on the top edge and one on the bottom edge. The cloth is then passed around the joint as far as possible. Then the cloth is taken in the left hand, with the fingers spread, and passed around the rest of the joint. If the solder does not take the shape of the cloth readily, then the solder is not at the right heat. This joint should be wiped very slim to allow room for the caulking tools. When this joint is once started, it should not be left until it has been wiped, otherwise a [44] large amount of solder will accumulate on the joint and will be hard to get off.

Points to Remember.—

1. *First*, material.
2. *Second*, tools.
3. *Third*, tin ferrule.
4. *Fourth*, use the dresser to fit the lead into the ferrule.
5. *Fifth*, clean the lead with the shave hook, and paper.
6. *Sixth*, use the catch cloth and ladle.
7. *Seventh*, keep the top edge covered with hot solder.
8. *Eighth*, wipe the top edge first.
9. *Ninth*, make a slim joint.
10. *Tenth*, steady the work with the free hand.

STOP COCK

Materials Required.—The materials used for this joint are as follows: two pieces of 5/8-inch extra strong lead pipe 9 inches long, each; one 1/2-inch plug stop cock for lead pipe; paste and paper; solder; 1/2 and 1/2 solder; rosin; catch pan and supports.

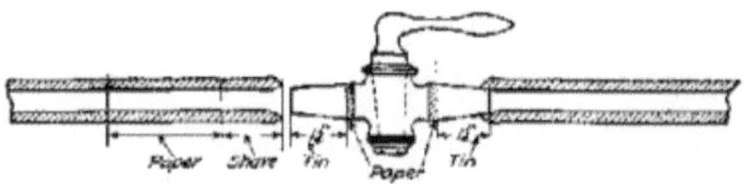

Fig. 28.

Tools Necessary.—The tools necessary for this job are as follows: saw, rasp, file, turn plug, shave hook, bending irons, hammer, ladle, soldering iron, and wiping cloths.

Preparation.—There are two joints to be wiped on this job and the stop cock is supported only by the rigid fitting of the lead pipe. Therefore the preparation must be thoroughly done. The brass stop will be prepared first.

Brass.—The two ends of the stop cock are filed bright, [45] then papered and tinned. This operation is the same, only on a smaller scale, as the tinning of the 2-inch and the 4-inch brass ferrule. The paper is pasted over the entire stop cock, except the two ends, which are tinned for about 1 1/4 inches.

Lead Pipe.—After the lead pipe has been cut off from the coil, the ends are squared with the rasp. One end of each piece is reamed out a little with the tap borer and spread a trifle with the turn pin. With the rasp, take off the outside edge of the end that has been spread. The sketch will show this and give the angle at which the edge is to be rasped. The stop cock is now fitted into the lead pipe. The brass should enter at least 1/4 inch, then the lead is beaten against the brass until a tight joint is made. The other end of the brass stop is fitted into the other piece of the lead pipe and a perfect fit is made. The fitting of these two joints must be rigid as upon them depends the stability of the joint support. When these ends of the lead pipe have been fitted, the pipe is cleaned with the shave hook and paper is pasted on, allowing 1 1/2 inches for the joint. Both pieces of pipe are prepared at the same time as both ends are wiped at the same time.

Supporting.—The three pieces of pipe should be so wedged together that they will not fall apart when put in position for wiping. The bricks for supporting the pipe are placed the same as in the support of the horizontal round joint. The lead pipe ends are laid on the bricks. This brings the stop cock in the center without any support. If it were not for the substantial fit between it and the lead pipe, it would not stay in place. Solder straps can be put over each end of the lead pipe. Weights can be used to advantage.

Wiping.—When getting the heat up for these joints, pour the solder over the two joints and over the stop cock. This gets the heat properly distributed, so that both joints [46] can be wiped while the brass stop is heated. Get the proper heat up on one joint and then the other. Come back to the first joint and wipe it and then the second one. Both joints should be wiped so as to have the same shape. The novice will experience some trouble when wiping this joint in getting the brass edge hot. Heating up the two joints together will in a large degree offset this trouble. Some mechanics take out the lever handle stop to lessen the amount of brass to heat. This is never done by a good mechanic as the two pieces will never fit together again and make a tight joint. If the plug is left in place, both the plug and body will expand equally and the pieces will fit perfectly. When wiping is started on these joints, the beginner must stay at it continually. When the brass is heated, the finished wiping can be tried over and over again. If this way is not followed, the beginner will find that most of his time will be spent trying to get a heat on the brass.

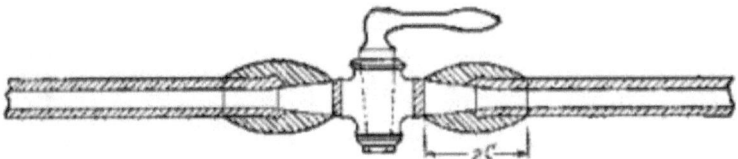

Fig. 29.—Stop cock.

BRANCH JOINT

Materials Needed.—The materials necessary to complete this job are as follows: 12 inches of 5/8-inch extra strong lead pipe for the run; 6 inches of 1/2-inch extra strong lead pipe for the branch; paste and paper, and solder.

Tools Necessary.—The tools necessary for this job are the saw, bending irons, rasp, tap borer, ladle, wiping cloths, and the shave hook.

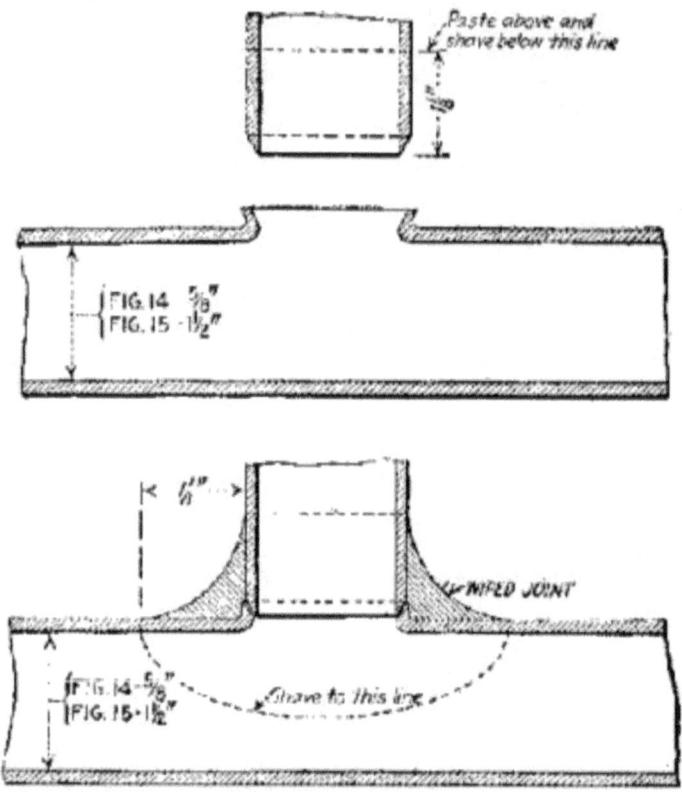

Fig. 30.—Branch joint.

Preparation.—The preparation of this joint requires [47] the skill of the beginner more than any of the preceding joints. The tapping of the 5/8 pipe for the branch connection, pasting and cutting the paper, require the utmost care and precision. The 5/8-inch pipe is tapped with the tap borer in the center. The tap borer is used by grasping the handle firmly and putting the cutting point on the mark and then pressing down on the handle. This forces the point into the lead. Now turn the tool and a piece of lead will be bored out. Continue this operation and a hole will very soon appear in the lead. A hole just large enough to allow the bending irons to enter is made. The opening of the hole is completed with the [48] bending iron, working the lead back slowly into place. Do not attempt to drive the lead back around the hole with a few strokes. One bending iron is inserted and this iron is struck with another iron or hammer. After a number of strokes the opening will be of sufficient size. The bent end of iron is inserted into the hole and the bent part enters the bore of the pipe. This iron is struck in such a way as to force the lead around the hole up, rather than back. Now with the straight end of irons open the sides. When the wall of pipe has been driven up a little the hole can be enlarged by driving back the lead. This procedure will form a collar around the hole to steady the branch pipe. Good workmanship will result in having a good substantial collar around the opening. The branch should now be fitted. Clean the pipe with the shave hook for about 2 inches on each side of the opening. With compasses set at 1 1/8 inches, mark off a space on each side of the branch on the run, or on the 5/8-inch pipe. On the sides of the pipe the two lines should be joined with an even and symmetrical curve. A good way to make this curve is with the shave hook. Now take a folded piece of paper and cut out the shape of one-half of the joint, then open the fold and the entire ellipse will be made. When this paper is cut, a sharp knife is used, otherwise a ragged edge will be made and a good finish of joint is impossible. The paper is now pasted and put on the pipe. The surplus paste on the edge of the paper should be wiped off with the fingers before the paper is put on the pipe. This prevents any paste squeezing out on the joint. The branch is now taken and perfectly fitted into the run. The end is cleaned with the shave hook and paper is pasted on the pipe, leaving 1 1/8 inches of cleaned surface for wiping. The paste and the paper should now be allowed to dry. The position for wiping this joint is to have the run horizontal and the branch on an angle of 45° pointing away from the wiper. Figure [49] 30 will bring out the above explanation very vividly.

Supporting.—The run of this joint is laid flat on the table and the branch inserted in its proper place. With one hand hold it in place, with the other, use the bending iron, tap the collar on the run against the branch, wedging it in place good and strong so that no solder can leak through. If the branch is tapered with the rasp as shown the joint can be made very tight. The run of the pipe is now laid on two bricks as was done with the horizontal joint. The branch is laid over on a pile of bricks or wood at an angle of 45°. The best way to secure this joint is to pour some half-molten solder on the ends of pipe and brick, making a solder clamp. This branch does not need any clamp or weight if it is properly entered into the run. A strap of solder can be run over the end of pipe if found necessary. Place the catch pan under the joint and then the pipe will be ready to wipe.

Wiping.—In wiping this joint, the catch cloth is used not only to catch the solder as it drops off from the pipe, but also to hold the hot solder against the pipe to heat the under side of the joint. Test the solder and see if it is the correct heat for wiping. If so, prepare for wiping. After heating the ladle, take some solder in it and proceed to drop the molten solder on the joint. The ladle is moved constantly as the solder is dropped on the run and then on the branch to get the entire joint to the proper heat. As the solder drops off from the joint, it is caught on the catch cloth and brought up on the top of the joint where it is re-melted by dropping hot solder on it. Then the hot solder is held in the cloth against the under side of the joint to get the under side properly heated. The solder is worked around all parts of the joint. When the heat is got up sufficiently and the solder works freely around the joint, the branch cloth is taken and each edge of the joint is wiped clean. Any surplus solder is brought up on top of the joint [50] and then wiped on the catch cloth. This solder is then put on the under side of the joint. With the branch cloth reach way around the joint and wipe each side, bringing the cloth each time to the top and then off the joint. The last wipe is directly across the top, wiping off any surplus solder that may have accumulated from wiping the sides. The difficulty with this joint is in getting the top and bottom to have an equal amount of solder. With a little practice and by watching each motion your faults can be noted and remedied. If the paper starts to come off, it should be re-papered at once. When the joint is finished, it should be left in position until the solder has had time to set and cool, otherwise the branch will break off and considerable time will be lost in correcting the trouble.

Points to Remember.—

1. *First*, the use of the tap borer.
2. *Second*, the use of the bending irons.
3. *Third*, do not allow the bending irons to touch the inside walls of the pipe when stretching the opening.
4. *Fourth*, secure the branch into the run.
5. *Fifth*, secure the pipes into position for wiping.
6. *Sixth*, spread the heat on the edges and the bottom of the joint.
7. *Seventh*, wipe with the branch cloth.
8. *Eighth*, cut the paper.
9. *Ninth*, mark the outline of the joint.

BRANCH JOINT PLACED FLAT

When the wiper has mastered the branch joint placed at an angle of 45°, he can proceed to wipe the joint placed in the next position, which is flat.

Preparation.—The preparation of this joint is identical with the preceding one placed at an angle of 45°. If a new joint is to be prepared, it would be well to pay strict attention [51] to the details, such as keeping the paste on the paper only and having the edge of the paper cut perfectly smooth and even. Before putting on the paper see that the pipe is free from all grease and dirt. The paste and paper will stick better if all the dirt is removed. The branch should be well fitted into the run of the pipe so that no solder will get into the bore of the pipe. The branch should not extend into the run of pipe enough to obstruct the bore of it. If the instructions for preparing the pipe are not carried out as detailed, the wiper will experience some trouble that he may find hard to overcome.

Supporting.—The run can be supported on bricks. The branch can be supported on a brick placed at its end the same height as the run. This will bring the joint in the correct flat position. The branch should point away from the wiper. Solder straps can now be poured over the ends of each pipe. If weights are used to hold the pipe firm instead of solder straps, they should be so placed that they will not interfere with the hands when wiping.

Wiping.—The wiping of this joint is more difficult as the beginner will experience trouble in heating the bottom and keeping the solder on the bottom. Solder is dropped on the joint and along the pipe so as to bring the pipe to the proper wiping heat. Some solder will accumulate on top of the joint. This is melted off on the catch cloth and this hot solder held against the bottom of the joint. This operation is repeated until the bottom as well as the top of the joint is heated properly. When the solder can be worked freely around the pipe, the branch cloth is taken and each side is wiped from the bottom toward the top. Solder is accumulated on the top where it is wiped off on the catch cloth and put on the bottom of the joint. Now reach way around each side and wipe the edge and body of the joint, a wipe across the top completing the joint. The bottom can be wiped with a cross wipe also if desired. The top and [52] the bottom should be identical. Notice carefully the drawing of this joint and endeavor to have the same lines. The perfecting of these joints comes only with patient practice. The beginner must not get discouraged because of a burn or two. As soon as confidence in oneself has been gained, the possibility of burning the fingers is entirely eliminated.

BRANCH VERTICAL

The materials, tools, and preparation for this joint placed in a vertical position are just the same, practically, as those in the preceding branch joints. One or two points wherein they differ are mentioned below. To rigidly support the joint for wiping, allow the run of the pipe to rest on some bricks as before mentioned, with the branch looking up. Now take a piece of wood and drive a nail through one end of it about 1 inch from the edge. Let this nail enter the bore of the vertical branch. The wood is allowed to rest on the back of the bench or is braced against the wall. Supporting the pipes in this way will allow the wiper perfect freedom. When wiping this joint, splash the solder on from the ladle as on the upright joint. As all the sides of this joint can be seen, it is not a difficult matter to make a perfectly symmetrical solder bulb. When the proper heat is gained, the top edge of the joint is wiped first, then the lower curved edge, using the branch cloth. The body of the joint is then wiped and the joint finished with a cross wipe, if necessary.

BRANCH HORIZONTAL

The next position for this joint is to have the branch pipe horizontal and the run vertical. The materials, tools and preparation for this joint are the same as for the preceding ones. The supporting and wiping differ a little.

Supporting.—One end of the run is placed on the catch [53] pan. The other end is held in place the same way as the branch was held in the preceding joint. If the pictures of this joint are carefully looked over, the methods employed to hold the pipe will be readily noted. The branch is best held by inserting one end of a bending iron in the bore of the pipe and placing the other end of the iron on a brick built up to the right height. The iron should be weighted to keep the joint from swaying.

Wiping.—The solder is now dropped on the branch as in the round joint, and splashed on the vertical run as in the upright joint. Sufficient solder is put on the joint to keep the edges covered with hot solder. Solder is worked around the joint until all parts of it are thoroughly heated and the solder works easily, then all the edges are wiped clean. The top half is then wiped evenly and the bottom half wiped to match the top half. A cross wipe in front completes the joint. When this cross wipe is made on any joint, a thick edge of solder must not be left. The edge must be wiped clean. This joint should be wiped first with the branch pointing to the right and then with the branch pointing to the left. It will take the beginner some time to master these branch joints, for not only must they be wiped symmetrically for the sake of appearances, but they must be wiped while the solder is hot to secure a tight joint. A joint that is wiped with solder that is too cold will be porous and will leak when put under pressure. With care the same pipe can be used throughout for all the positions of this branch joint.

ONE AND ONE-HALF-INCH BRANCH JOINT

Upon the completion of the small sized branch joint in its various angles, the 1 1/2-inch branch joint is to be wiped. This branch joint is wiped in the same positions as the 5/8 branch was wiped. The pipe being larger, there is more [54] solder for the wiper to handle, and the edges to keep clean and to wipe are longer.

Materials Needed.—The materials needed for this job are 12 inches of 1 1/2-inch light lead pipe for the run, and 6 inches of 1 1/2-inch pipe for the branch, paste, paper, solder, and catch pan.

Tools Needed.—The tools necessary for this job are the saw, rasp, shave hook, bending irons, drift plug, hammer, ladle, wiping cloths, and tap borer.

Preparation.—To an experienced wiper, the procedure of preparing this joint and wiping it are so near like the 5/8-branch joint that a detailed description would be unnecessary; but for the benefit of the beginner, I will repeat the details as they apply to this particular joint and thereby avoid any error. We will take the preparation of the run first. Square the two ends of the pipe with the rasp. Mark off the center of the pipe. With the round part of the rasp, held at right angles with the pipe, proceed to rasp down the crown of pipe where the center mark was made. Do not rasp through the wall of the pipe, but just enough so that the tap borer will enter the pipe with only a slight pressure. With the tap borer, tap a hole large enough for the bending irons to enter. Now proceed to enlarge the hole, first forcing the edges up and then forcing them back, making the hole larger and making a collar around the hole at the same time. Continue to open the pipe until the aperture is large enough for the branch pipe to enter. The bending irons must not come into contact with the inside wall of the pipe, for if they do the inside bore will be marred and be very ragged. As these joints are usually used on waste lines, these ragged places make an ideal place for lint and grease to collect and cause a stoppage. To make the inside of the hole even, a piece of 1/2-inch pipe can be used in place of the bending irons. To cut out the oval from a piece of paper to fit the joint, fold the paper and cut out one-half [55] of the oval. Now unfold the paper and the complete oval is obtained. The measurements of the oval are taken from Fig. 30, 1 1/8 inches each side of the branch lengthwise of the run. These two lines are connected with a curved line as shown. This curved line can be made with the shave hook. Take the large edge of the shave hook and roll it along between the lines to be joined. A little practice will perfect one in doing this quickly. The beginner should make a number of these ovals so that he can get them perfect. The graceful appearance of this joint depends upon the neatness with which it is prepared. I do not want the beginner to think that a graceful shape of the joint is all that is to be desired or that it is the most essential point. Further along, perhaps, more vital requirements will be brought out and the beginner will be made acquainted with them.

The ends of the 6-inch piece are now squared with the rasp. The edges of one end are rasped off as shown in the sketch, making a wedged fit into the run. This end is then cleaned with the shave hook. Paper is then pasted on to cover the pipe except the 1⅛ inches cleaned on the end. This cleaned part forms part of the joint, therefore no paste or paper must be put on it. The pipe is now fitted into the run and the collar beaten against it with the bending irons. The run is now cleaned with the shave hook for about 3 inches each side of the center. The paper oval cut out is now pasted on the joint. The paste and paper are then allowed to dry before they are handled further.

Supporting.—The supporting of this joint, which is placed with the branch on an angle of 45° pointing away from the wiper, is not a difficult matter. The beginner can use his own ingenuity for supporting the pipe if conditions do not warrant the using of the methods previously described.

Wiping.—The solder should now be tested for heat. If [56] the solder is at the proper heat, the ladle is taken and heated. Take a ladle full of solder and drop the solder on the joint. The lead of which this branch joint is made is considerably lighter than any lead that has been used before. Therefore, the beginner must drop the solder on carefully, making sure that the solder is not dropped on the same spot, for a hole can be burned through the pipe very quickly. The ladle must be kept moving, then the solder will not burn through the pipe. The heat is got up on the pipe by dropping the solder on the run and on the branch, catching the surplus solder on the catch cloth and heating the under side of the joint with it. To form the joint, distribute the solder and then wipe it into shape. Notice that I said wipe it into shape. A beginner is very apt to try to push or poke it into shape. This must not be done as it has a tendency to make the joint lumpy. All the edges are wiped off clean first, then the body of the joint is shaped and wiped. When forming the joint, be sure that the bottom and the top are symmetrical. Do not have one-half larger than the other. The last wiping strokes are made swiftly and rapidly. If the wiper will watch his movements and note the results and then try to improve them, keeping in mind that a symmetrical joint is wanted with thin edges, perfection in wiping will come much more quickly than if no attention is paid to the strokes made when wiping.

BRANCH JOINT WIPED FLAT

The materials required for this joint do not differ from the preceding one. If the pipe used for the branch joint at a 45° angle is in good shape, it can be used for this joint by simply changing positions. The tools needed will not be any different. The ladle and the wiping cloths, of course will be required. A pair of pliers can be used to advantage in picking up the hot solder. The wiping cloths [57] should receive a little more oil to keep them soft and pliable. Oil the edges of the cloths well.

Supporting.—To support this pipe for wiping have each end rest on a brick. Each end can be weighted to hold it in place.

Wiping.—To wipe this joint, proceed to drop the solder on the joint. When the pipe is thoroughly heated and the solder works freely around the pipe the joint can be wiped. The procedure is like the preceding one. The wiper is cautioned to move the ladle constantly while dropping the solder.

BRANCH HELD VERTICAL

After a number of the previous joints have been wiped successfully, the pipe is placed in such a position that the branch will be vertical. The supporting of the pipe to hold the joint in this position for wiping is very easily done after handling the 5/8-in. joint in this position. The following points may be found helpful: The solder is splashed on the joint from the ladle. The top edge of the joint is kept hot by keeping the solder covering it. When the proper heat has been got up, the top edge is wiped first, then the bottom edges both front and back. The body of the joint is wiped last and a cross wipe finishes the joint. I have found that the beginner in many cases, when this joint is reached, tries to wipe it with many short strokes. The habit is a bad one and should be stopped as soon as noticed.

Learn to wipe the top edge with only two strokes, the bottom edge with not more than four, the body of the joint with four, and one cross wipe to finish. This joint should be finished as symmetrically as possible and wiped while the solder is hot.

[58]

RUN HELD VERTICALLY

When the vertical branch has been conquered and the wiper can get a good joint every time it is tried, the pipe can be changed to a different position. The run is placed in a vertical position and the branch horizontally to the left. The catch pan is put under the end of the pipe. Follow the same directions for supporting this joint as were given under the 5/8-in. branch placed in a similar position. The wiping of this joint is so nearly like the preceding branch joints that I will not give any instructions at all. This joint is finished at the same point that the other branch joints are finished. However, there are one or two matters that should be kept in mind. Some of the small matters are often overlooked and should be called to mind occasionally. Do not allow the solder to accumulate in the pan. If the cloths are burned, they should be turned, or new ones made. If the paper has started to come off from the pipe, new paper should be put on at once. Test the solder occasionally and see that it does not get too hot. Upon completion of the joint in this position, the branch joint in its various positions is finished. The beginner has found out while wiping these various joints a number of points that were not mentioned in my description. No amount of detailed description will make a good joint wiper. Patience and practice are as important in joint wiping as good preparation and good solder.

Points to Remember. —

1. *First*, materials — 18 inches of 11/2-in. lead pipe.
2. *Second*, use of tools.
3. *Third*, keep bending irons away from the wall of the pipe.
4. *Fourth*, make a good collar around the opening.
5. *Fifth*, make a tight fit with branch and run.
6. *Sixth*, hot solder will quickly burn through the lead.
7. *Seventh*, use branch cloth for wiping.

8. *Eighth*, cut out paper for joint even and symmetrical.

[59]

BIB

This joint is another brass to lead, and is the last single joint to be wiped in this course of joint wiping.

Materials Needed.—The materials required for this joint are as follows: 10 inches of 5/8-inch extra strong lead pipe; one 1/2-inch brass sink bib for lead pipe; one pot of solder, paste and paper, 1/2 and 1/2 solder, catch pan, and supports.

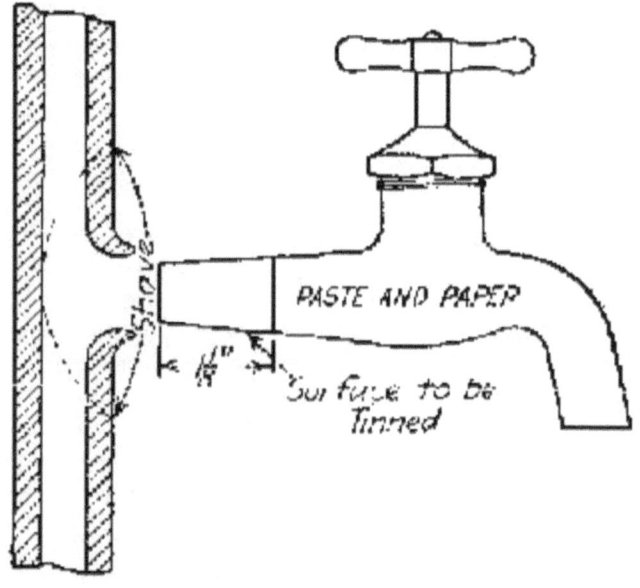

Fig. 31.

Tools Required.—The tools required for this job are the saw, rasp, tap borer, bending irons, file, ladle, wiping cloths, shave hook, knife and rule, soldering iron.

Preparation.—To prepare the lead pipe after cutting from the coil and squaring the ends with the rasp is very similar to the 5/8-inch branch joint. The center of the pipe is marked and a hole is made in it with the tap borer large enough to admit the bending irons. The hole is enlarged with the irons. A good substantial collar is made around the hole to hold the bib in place. One and one-eighth inches are marked off on each side of the branch and an easy curve connects the two. The paper is then cut out and pasted on the pipe after it has been scraped with the shave hook.

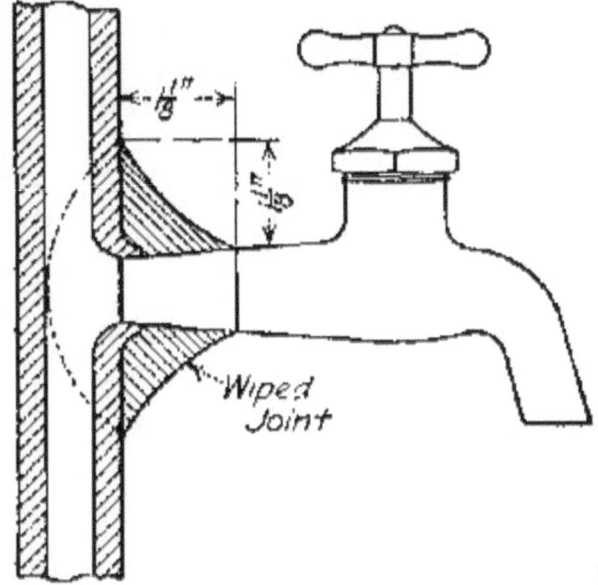

Fig. 32.—Bib.

The end of the brass bib is filed bright and tinned with the soldering iron and 1/2 and 1/2 solder. Before the tinning is done, paper is put on the brass, leaving only 1 1/8 inches exposed. The tinning must be thoroughly done, or it will come off and have to be re-tinned.

Supporting.—The bib is fitted into the lead opening and the collar is forced against the bib to hold it in place and prevent any solder from leaking through into the bore of the pipe. The bib must not extend too far into the lead pipe [60] or it will obstruct the flow of water. The lead pipe is laid on two bricks the same as the round joint. The bib is laid on an angle of 45° pointing away from the wiper. Some bricks can be piled up to the right height to hold the bib in place and a solder strap can be made to hold it steady. The lead pipe can be held steady by weighting each end. The catch pan is now placed under the joint and everything is ready for wiping.

Wiping.—When the solder is hot, getting the heat on the pipe is started. Solder should be dropped oftener on the brass bib than on the lead pipe. It takes more heat to heat the brass thoroughly than it does the lead. If this is followed out, little difficulty will be had in getting up the heat and in wiping. Use the branch cloth for wiping and make sure that all edges are perfectly cleaned before making the final strokes. As this is the only position that the joint will be wiped in, practice should be continued until perfect joints can be obtained.

Points to Remember.—

1. *First*, materials needed.
2. *Second*, tools needed.
3. *Third*, use tap borer.
4. *Fourth*, enlarge hole with bending irons.
5. *Fifth*, make substantial collar around the opening.
6. *Sixth*, paper the lead.
7. *Seventh*, file the bib, then paper.
8. *Eighth*, tin the bib.
9. *Ninth*, place in position and wipe.

[61]

DRUM TRAP

The making of the drum trap will bring out the skill of the beginner. The entire trap is made of lead pipe. The lead will require a great deal of handling. Therefore, care must be exercised in all operations to turn the trap out in a workmanlike manner.

Materials Needed.—The materials needed to complete this job are: 10 inches of 4-inch 8-pound lead pipe; 18 inches of 1½-inch light lead pipe; paste and paper, support, solder, and catch pan.

Tools Needed.—The tools required for this job are: saw, rasp, bending irons, shave hook, bending spring, tap borer, dresser, ladle, drift plug, and wiping cloths.

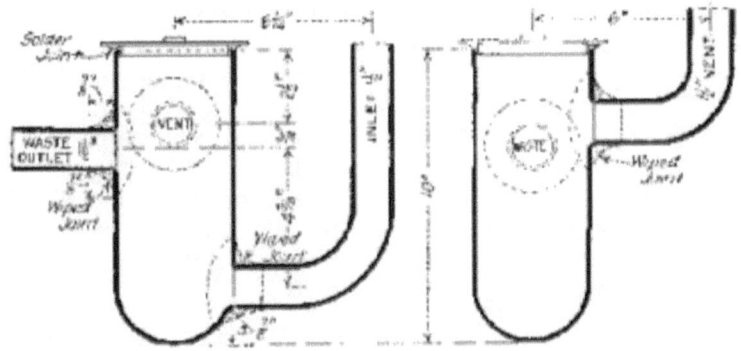

Fig. 33.—Drum trap.

Preparing.—Take the 10-inch piece of lead pipe and hold it in one hand, in the other hand take a pine dresser. Strike the lead pipe with the dresser. The pipe is struck about 2 inches from the end and is beaten evenly all around. The pipe is then struck nearer the end until finally the bore of the pipe is almost closed. This closed end should be rounding and symmetrical. To get this shape the pipe must be continually moved and turned. One side must not be forced in more than the other. If there are any dents in the pipe or part of the pipe is forced in too much it may [62] be driven out as follows: Take an old piece of 1/2-inch lead pipe and round one end of it with a hammer; this can be used by hitting the inside of the closed end of the drum and forcing out the dents. The rounded end of the trap is not quite closed and a hole about 3/4 inch is left. This opening is closed by shaping the edges of it with the knife, making them smooth and beveled. Then a piece of lead is cut out of some scrap, the same shape as the hole and fitted into it. The top surface of this fitted piece should be a little lower than the surface of the pipe. Strike a circle, using the compasses, the center of the circle being the center of the inserted piece of lead. The lead inside of this circle is shaved clean with the shave hook, including the inserted piece. Paper is then pasted outside of the circle and should cover entirely the rest of the pipe. The inserted piece is wiped on the pipe as follows:

Wiping End.—Stand the 4-in. pipe in a pan with the rounded end of the pipe up. Be sure that the inserted piece is fitted securely. The solder is now dropped on the paper and shaved portion of the pipe. Exercise considerable care not to burn a hole in the pipe. As the hot solder runs off, catch some of it and draw it back on the joint. When the solder can be manipulated freely and the pipe is hot, the joint can be wiped. The cloth is drawn across the joint, cleaning all the edges with one stroke. The joint should be shaped to complete the rounding surface of the pipe. The joint is comparatively easy and will not occupy much time. As soon as it is wiped, cover the solder with paper. This will preserve the freshness of the joint until all wiping is completed.

PREPARING INLET PIPE

After the above joint is completed, the 11/2-in. branch inlet pipe is prepared and wiped in place. The center of this branch is marked on the 4-inch pipe and a hole is tapped [63] in the pipe, using the tap borer. A hole large enough to admit the bending irons is made. The hole is enlarged with the bending irons, bending the lead first *up*, then *back*. A piece of 1/2-inch iron pipe can be used as a tool to finish the opening. The iron pipe is larger in diameter than the bending irons and leaves a more finished surface. The opening is made of sufficient size to admit the rasped end of the 11/2-inch pipe. When using the irons to enlarge the opening in the pipe, be sure not to bruise any part of the trap. The 11/2-inch pipe is now taken. The ends of this pipe are squared with the rasp. The drift plug is then driven through the pipe to take out any bruises or flattened places. The edge of one end is rasped off to fit the opening made in the 4-inch pipe. The beginner must strive to make a perfect fit. The accuracy with which these preparations are made is what helps in a large degree to bring about a successful job. The next operation is to paper the parts not to be wiped. The sizes of the joint should be followed as shown on the sketch. The pipe is first shaved with the shave hook, after which the paper is pasted on. No paste is allowed to get on the joint proper. The beginner should by this time have formed the habit of being neat with his work. Therefore the getting of paste on the joint surface shows that he is not as neat or as far advanced as he should be.

Supporting.—The drum is laid lengthwise on the bench and blocks are put on each side to keep it from rolling, the branch uppermost. The 1½-inch pipe is held in position the same way as the vertical branch was held. The catch pan is put under the drum to catch the surplus solder.

Wiping.—Splash the solder on the branch pipe, also on the drum. The burning through of the drum is an easy matter. Therefore do not keep dropping the solder on one place, but keep the ladle moving continually. With the catch cloth draw the solder up on the branch covering the [64] top edge of the prepared surface. Splashing the solder on this top edge melts the solder already on and allows it to run down on the 4-inch pipe where it is caught with the cloth and again brought up on the top edge of the branch. When the solder works freely all around the joint, the top edge is wiped clean and even. Then any surplus solder is wiped off. The bottom edge is next wiped clean, after which the body of the joint is wiped into shape, together with both edges. The edges are wiped very thin so that when the paper is removed the outline of the joint stands out very distinctly. A thick edge on a joint gives an unworkmanlike appearance to the work. The joint is finished with a cross wipe.

The other joints are prepared and wiped the same as the one just completed. The 1½-inch branch connection taken out of the bottom of the trap is bent. As this is the first time it has been necessary to bend lead pipe in these jobs, I will cover this operation in detail. The pipe is first straightened and the drift plug driven through it. The pipe is marked where the bend is to be made. The bending spring, size 1½ inches, is put into the pipe, the center of the spring coming about where the bend is to be made. The pipe is then heated where it was marked to be bent. The proper heat for this pipe is just so that the hand cannot stand being laid against it. The pipe is held in the hands and on the end nearest the heat is hit against the floor at an angle. The pipe, with the first blow, will start to bend. With a few more strokes the desired bend will be obtained. The bending spring can now be pulled out. Put a little water in the pipe, then put one end of the spring in the vise, twist the pipe, and the spring will come out when the pipe is pulled away from it. The bending spring holds the pipe cylindrical while it is being bent. Without the spring, the pipe would be badly crushed at the bend and rendered almost unfit for service. Another good way [65] to bend pipe is to plug one end and fill the pipe full of sand, then plug the open end. The pipe is then heated where the bend is to be made. The pipe can then be bent over the knee. When all the joints are wiped, the paper should be taken off and the lead cleaned with sand and water. The trap is now complete except the brass clean-out to be soldered on the top. The inside of the trap should not have any rough edges or drops of solder in it.

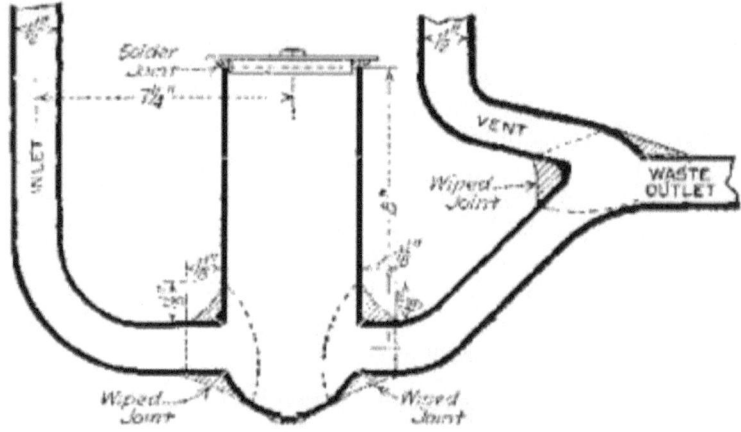

Fig. 34.—Drum trap.

There are two other drum traps to be made. The materials needed are the same as for the above trap except for 18 inches more of 1½-inch lead pipe. The support, preparation, and wiping are the same. The beginner by this time should feel very well acquainted with lead and solder. Therefore, the details of these two drum traps can be left for the beginner to work out for himself. The sketches are very distinct and readable and will be of considerable assistance. The beginner should make these traps.

Points to be Remembered.—

1. *First,* use 4-inch lead pipe, 8 pounds to the foot.
2. *Second,* dresser and spring are new tools. Study their use.
3. *Third,* gradually work the trap into shape with the dresser. [66]
4. *Fourth,* plug the hole with a piece of lead pipe.
5. *Fifth,* prepare and wipe the plugged hole first.

6. *Sixth*, prepare and wipe the 1½-inch branches.
7. *Seventh*, special care should be taken to keep the work neat.
8. *Eighth*, two ways of using the bending spring.
9. *Ninth*, wipe thin edges on joints.
10. *Tenth*, do not handle finished work.
11. *Eleventh*, clean and finish the work neatly.

THE PRACTICAL USE OF THE PRECEDING EXERCISES

In the foregoing exercises, I have confined myself to the actual work of making the various joints. Now I will explain the practical use of them.

Soldering Iron.—The soldering iron is a tool that is used in work that requires heat to fuse solder and the parts to be united. Every plumber should have at least two irons in his kit.

The Cup Joint.—While the cup joint is not employed to any great extent in modern plumbing, yet it has its use in the installation of some fixtures. Lavatories, bath and toilets are sometimes connected with a short piece of lead on the supply. The tail pieces on the faucets can be soldered on the lead by means of a cup joint. A cup joint well made with a deep cup and the solder well fused is as strong as a wiped joint in a place of this kind. The evil of the cup joint is that some mechanics will only fuse the surface and leave the deep cup only filled with solder and not fused. This makes a tight joint, but extremely weak. On tin-lined pipe and block-tin pipe the cup joint is commonly used. When making a cup joint on block-tin pipe the soldering iron must not touch the pipe and fine solder should be used. When tin-lined pipe is being soldered, the tin lining must not be melted.

Overcast Joint.—The overcast joint is not commonly [67] used, but when there is considerable lead work to do the plumber finds it very handy in places where a wiped joint would take up too much room. We use it for an exercise for the reason that it teaches the beginner very rapidly the use and control of the soldering iron.

Flat Seams.—These seams are used in the construction of roof flashers, tanks (Sec. 33, Chapter XVIII) and lead safe wastes (Sec. 27, plumbing code). A hatchet iron is sometimes used on these seams.

Wiping Cloths.—The wiping cloths made of whalebone ticking make good, serviceable, and lasting cloths. Oil only should be used to break the cloth in. Moleskin cloths are very good, but they are very hard to get and cost considerably more. A plumber should always keep a good supply of ticking cloths on hand. The cloths are used only for wiping.

1/2-inch Round Joint.—This joint is the one most often required in actual practice. It serves to connect two pieces of lead pipe of the same or different diameters. It is also used to connect lead and other materials of which pipe is made. The workman, when he gets out on the job, finds that his work cannot be supported for wiping in such an easy and convenient position as illustrated in the exercises. It will be necessary to wipe the joint at almost every conceivable angle and position. The workman must employ his ingenuity to overcome any difficulties that may arise. Any draught of air should be avoided as it will make the solder cool quickly.

2-inch Brass Ferrule.—When it is found necessary to connect cast-iron and lead pipe, it is done by means of a brass ferrule wiped on the lead pipe. This joint is a very common joint and is found on sink, tray, and bath connections, as well as in many other connections that have lead and cast-iron pipes for wastes.

4-inch Brass Ferrule.—The 4-inch brass ferrule wiped [68] on lead pipe is found under almost every closet. There is generally a piece of lead connecting the toilet with the soil pipe. Therefore, a brass ferrule is wiped on the lead and the ferrule connected with the soil pipe. This joint is also found on rain leader connections near the roof, connecting the gutter with the rain leader stack.

Stop Cock.—When a shut-off is required in a line of lead water pipe, these joints are used. Where it is necessary to joint lead and brass, this joint is required. The art of heat control over the lead and the brass is the essential point in these joints.

Branch Joints 5/8 and 1/2 Inches.—Where it is found necessary to take a branch from a water pipe, this joint is used at the connection.

In practice, this joint may have to be wiped in positions that are rather difficult to reach, so the wiping of joints in the positions called for in the exercises is exceedingly good practice.

Branch Joints 1½ Inches.—These joints are very common and are found on waste and vent pipes. They are also found on urinal flush-pipe connections where the branch often is brass and the run lead.

Bib.—When lead supplies are run directly to the bib on a sink, this joint is necessary. It becomes necessary to wipe in a piece of brass for a brass-pipe connection from a lead pipe, in which case this joint is called for.

The Drum Trap.—The drum trap is used under sinks, baths, showers, and trays.

[69]

CHAPTER VII

Laying Terra-cotta and Making Connections to
Public Sewers. Water Connections To
Mains in Streets

TERRA-COTTA PIPE

One of the first pieces of work which a plumber is called upon to do, when building operations commence, is to run in the terra-cotta sewer from the street sewer into the foundation wall.

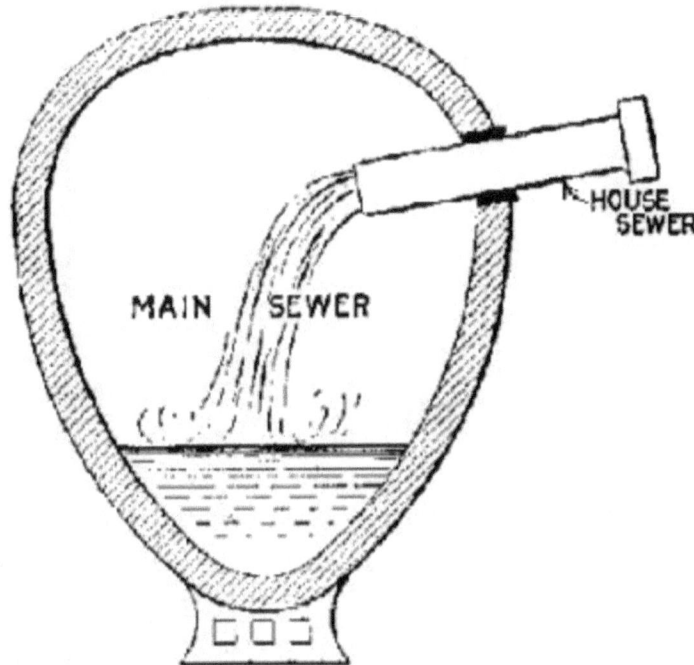

Fig. 35.—Connection of house sewer to main sewer.

When the street sewer is laid, Y-branches are left every few feet. A record of the branches and their distance from the manhole is kept generally in the Department of Sewers or Public Works. Therefore, the exact measurement of any branch can be obtained and the branch found by digging down to the depth of the sewer. A branch should be chosen [70] so that the pipe can be laid with a pitch, the same way as the main sewer pitches. This can be done by getting the measurements of two of these branches and choosing the one that will serve best. When there is a brick sewer in the street and no branches left out, the sewer must be tapped wherever the house sewer requires it (see Fig. 35).

Digging Trenches.—After the measurements and location of the house sewer and sewer branches are properly located, the digging of the trench is started. The methods employed to dig the trench vary according to the nature of the ground, that is, whether it is sand, rock, or wet ground. A line should be struck from sewer to foundation wall to insure a straight trench.

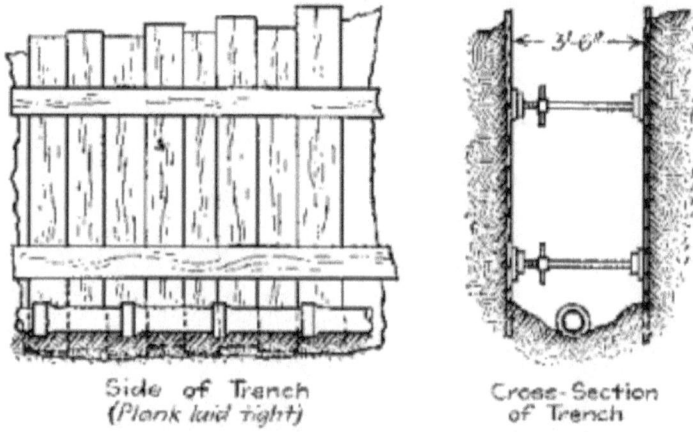

Fig. 36.—Laying of plank for trench dug in sandy ground.

Sandy Ground.—If the ground is sandy, the sides of the trench will have to be sheathed or planked and the planks braced so as to prevent the bank caving in. As the trench is dug deeper, the planks are driven down. When the trench is very deep, a second row of planking is necessary. The planks must be kept well down to the bottom of the trench and close together, otherwise the sand will run in. It is well to test the planking as progress is made by tamping [71] the sand on the bank side of the planks.

Gravel.—Where the ground is mostly gravel and well packed, the above method of planking is unnecessary. The bank should have a few stringers and braces to support it. When only a few planks are used the term "corduroy the bank" is used (see Fig. 37).

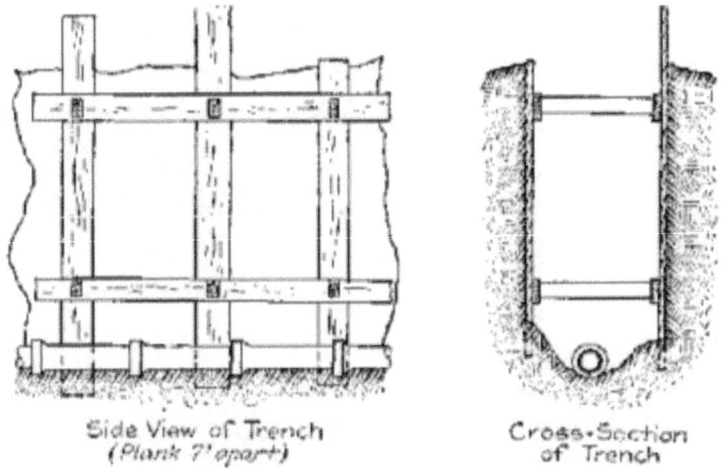

Side View of Trench
(Plank 7' apart)

Cross-Section
of Trench

Fig. 37.—Arrangement of plank for gravel.

Rock.—Where rock is encountered, blasting is resorted to. The plumber should not attempt to handle a job requiring the use of powder. It is dangerous in the hands of a person not used to handling it and the work should be sublet.

A sketch of the two methods above for planking trenches is given and a little study will make them clear.

LAYING OF PIPE

The pipe should be laid on the bottom of the trench to a pitch of at least 1/4 inch per foot fall. In laying, the start should be made at the street sewer with hubs of pipe toward the building. The trench should be dug within a few inches of the bottom of the pipe, then as the pipe is [72] laid the exact depth is dug out, the surplus dirt being thrown on the pipe already laid. The body length of pipe should be on solid foundation. A space dug out for each hub as shown in Fig. 38 allows for this, also allows for the proper cementing of joints. To get the proper pitch of pipe, take for example 1/4 inch per foot, a level 2 feet long with a piece of wood or metal on one end 1/2 inch thick will answer. The end with the 1/2-inch piece on should be on the lower hub and the other end resting on the hub of the pipe about to be put in place. When the bubble shows level, then the pipe has the 1/4-inch fall per foot. If a tile trap is used, it should be laid level, otherwise the seal will be weakened or entirely broken.

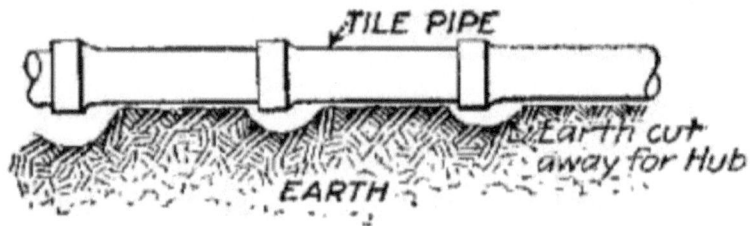

Fig. 38.—Laying terra-cotta pipe.

Cutting.—The cutting of tile is not difficult, but must be done carefully or the pipe will crack or a piece will be broken out, thus making the pipe worthless. To cut tile or terra-cotta pipe, stand the pipe on end with the hub down, fill the pipe with sand to the point of cutting. With a sharp chisel and hammer cut around the pipe two or three times and the pipe will crack around practically straight.

Cementing.—If the pipe is free from cracks, the only possible way roots can get into the inside of terra-cotta pipe is through the cement joint. There are two ways of making these joints. Both ways are explained below and are used today on terra-cotta work.

> 1. *First.*—The bottom of the hub of pipe in place is filled with cement and the straight end of the next piece of pipe is laid in place, then more cement is placed into the hub until [73] the space between the hub and the pipe is filled. In a trench, a trowel is rather unhandy to work with, while the hands can be used to better advantage. The cement can be forced into place with the hands and then surfaced with a trowel. The rest of the operation is to swab out the inside joint to remove any cement that perchance was forced through the joint (see Fig. 39). The cement used should be 1/2 cement and 1/2 clean sharp sand.

Second.—Half of the space between the hub and the pipe is first packed with oakum and then the other half filled with cement of the same proportions as that used above.

Fig. 39.—Showing use of the swab.

LAYING PIPE IN TUNNELS

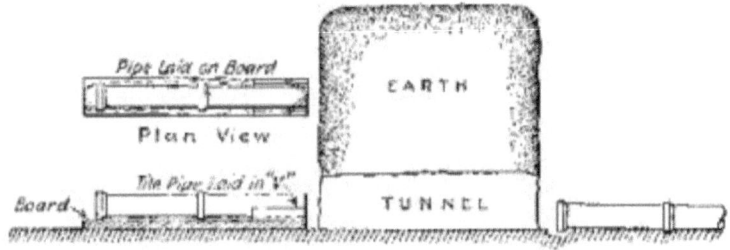

Fig. 40.—Pushing pipe through tunnel.

If the pipe must be run through a tunnel and there are perhaps three or four joints that cannot be reached, they should be put into place as follows: The pipe should be laid in the trench from the sewer in the street as far as the tunnel, then start at the other end of the tunnel. Lay the first piece of pipe on a board, lengthwise with the board, nail two cleats in the shape of a > (Fig. 40) for the pipe to rest in; push this pipe and board into the tunnel and then [74] cement into its hub a second piece; push the two pieces in 2 feet, cement a third length into the second piece and push the three pieces along 2 feet. A workman can be on the sewer side of the tunnel and receive the end of the pipe as it is pushed through the tunnel, and steer the pipe into the hub. The joints in the tunnel will not be as secure as those outside. This explains how pipe is run through a tunnel.

Connecting.—The proper method of connecting the house sewer with the street sewer is shown in Fig. 35. The connection should be made above the spring of the arch. The pipe should extend well into the sewer so the sewage will discharge into water and not drop on sides.

Inserting.—To insert a tee in a line of pipe already laid, pursue the following method (see Fig. 41): Cut or break out one joint, preserve the bottom of the hub of pipe that is in. Cut away the top of the hub on the pipe to be inserted, then place the pipe in position and turn around until the part of the hub on the piece inserted is on the bottom. The bottom part of the pipes now will have a hub to receive the cement. The top part will have to be cemented carefully, as it is within easy access. This can be done without difficulty.

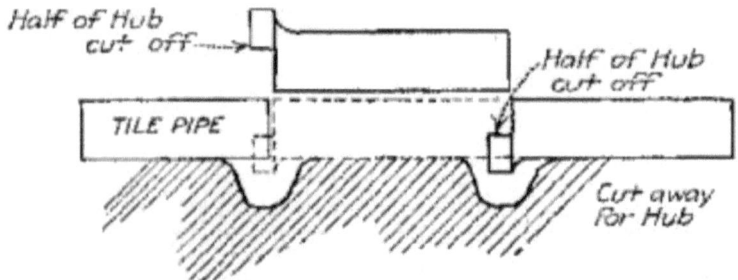

Fig. 41.—Inserting length of pipe.

While laying the pipe a stopper is used to prevent the sewer gases and foul odors from escaping. This stopper sometimes is of tile, sometimes a plug of paper or burlap. This stopper is sometimes cemented in by inexperienced [75] men and the trouble created can only be guessed at. If a stopper is used, the workman must see that it is taken out.

Refilling.—After the pipe is laid and cemented, it should be covered and allowed to stand 24 hours to give the cement time to harden. The dirt should then be thrown in and settled by means of a tamper or by flooding with water. The planks should not be taken out until the trench is well filled. To pull the plank, a chain or shoe and lever will have to be used. Where the tunnels are, dirt will have to be rammed in with a long rammer, care being taken not to disturb the pipe. If the refill is not well rammed and tamped, the trench will settle and cause a bad depression in the street surface.

Terra-cotta Pipe.—Terra-cotta pipe should be straight, free from fire cracks, and salt-glazed. The inside of the hub and outside of the plain end should not be glazed. This allows the cement to take hold.

Table of Standard Terra-cotta Pipe

Size	Thickness, inches	Weight per ft., pounds	Depth of socket	Annular space
3	1/2	7	1 1/2	1/4
4	1/2	9	1 5/8	3/8
5	5/8	12	1 3/4	3/8
6	5/8	15	1 7/8	3/8
8	3/4	23	2	3/8
9	13/16	23	2	3/8
10	7/8	35	2 1/8	3/8
12	1	45	2 1/4	1/2
15	1 1/8	60	2 1/2	1/2
18	1 1/4	85	2 3/4	1/2
20	1 3/8	100	3	1/2

Terra-cotta pipe should not be permitted in filled-in ground.

Roots of trees find their way into the pipe through cracks [76] or cement joints. When the roots get inside of the pipe they grow until the pipe is stopped up. As the roots cannot be forced or wired out, the sewer must be relaid. The writer has seen a solid mass of roots 10 feet long taken out of a tile sewer.

In case terra-cotta is laid in filled-in ground, there is only one way to insure the pipe from breaking. The pipe should be laid on planks. Then, if the ground settles, the pipe will not be broken.

WATER CONNECTION AND SERVICE

Tapping Main.—The water service for a building is put in at the same time as the sewer is connected and run into the house. For a 1¼-service pipe a ½-inch tap is furnished. The water company taps the main, at the expense of the plumber, and inserts a corporation cock.

Fig. 42.— Showing water main and sewer in same ditch.

Digging Trench.—The trench for the water main should be dug at least 4½ feet deep or below frost level and the trench should be kept straight. When the sewer is put in at the same time, one side of the sewer trench can be cut [77] out after it is filled up to the level of the water main. The water pipe can then be laid on this shelf at least 2 feet away from the original trench of sewer. Sometimes the surface of the ground must not be disturbed. In this case small holes are dug and the pipe is pushed through or driven through under that portion not dug. These places are often tunnelled (see Fig. 42).

In digging in city streets, care should be taken not to destroy any of the numerous pipes encountered.

LAYING PIPE

The trench should be dug straight out from the house so the pipe can be laid and the main tapped straight out from the building. The water companies keep a record of these taps so that in case of trouble the street can be opened and the water shut off. In laying the water service, the pipe from the curb to the main should be laid first. This takes in all the pipe in the street. At the main there is a shut-off in the tap. Another stop with T or wheel handle must be placed just inside the curb line. This is called a curb cock (see Fig. 43). One trench either outside or inside of the curb should be at least 15 feet long so that a full length of pipe can be laid in the trench. It is generally impossible to open a trench the full length the pipe is to be run. A trench 10 feet long is dug, then 8 feet left, and another 10- or 8-foot trench is dug and the two are connected with a small tunnel and pipe pushed through. When the pipe has been put in place between the curb and main, the water is turned on and the pipe flushed out. The valve at the curb should now be shut off, and if there are any leaks they will show. The street part is now ready to fill in. At this point Fig. 43 should be studied. Note the piece of lead attached to the pipe and corporation cock. This piece of lead should be extra heavy and always laid in place the shape of the letter [78] S or goose neck. In case the street should settle, this piece of lead will allow for it. These "lead connections" or "goose necks" are made as follows: 3 ft. of 5/8 lead pipe; 1-inch brass solder nipple (wiped on); one brass corporation cock coupling (wiped on).

Laying Pipe.—This lead connection can be screwed on the pipe after the pipe is laid, then bent and coupled on the main with the coupling.

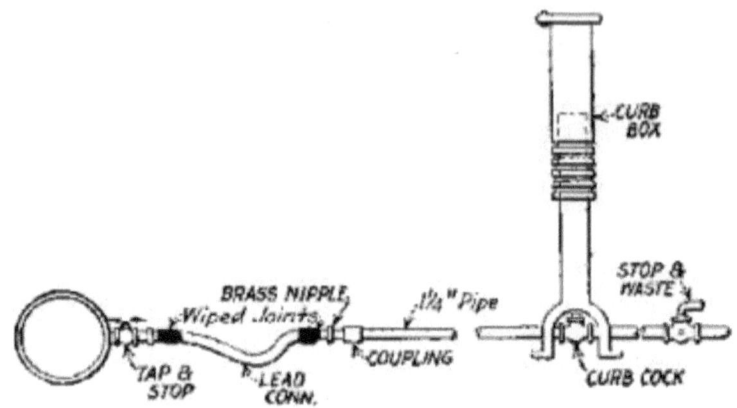

Fig. 43. — Water main from street to foundation wall.

After the pipe has been tested as far as the curb, the trench in the street can be filled as described later. The pipe from the curb to the building can now be laid. If necessary to push the pipe through a tunnel, the end of the pipe should first be capped. Start by screwing a length in the curb cock. If the other end of the pipe comes in a tunnel an additional length must be put on before putting in place so that an end will come in the open trench. When the building is reached and before the stop cock is put on, the valve at the curb should be opened full and the pipe flushed out. The valve can then be put on and water turned on to test the pipe.

Setting Curb Box. — A cast-iron box, adjustable length, with cover should extend from the curb cock to the surface. This makes it possible with a long rod to control [79] the water service into the building. To set a curb box some flat stones should be laid around the curb cock and the box set on these stones. Then the space around the box and pipe should be closed in with brick or other covering to keep the sand from washing in on the curb cock. The box should be adjusted for height and then held in place by placing the curb key rod in place and holding the rod and box while the trench is filled. The refill should be tamped evenly on all sides of the box.

Refill. — In refilling the trench around the corporation cock and goose neck, the greatest care should be taken. The writer has seen cases when indifferent workmen have tossed heavy stones in the ditch and broken off the corporation cock or destroyed the goose neck. After the pipe is covered with 18 inches of refill and tunnels have been filled, water can be run in the trench and will settle the refill.

There are a number of special points concerning water services and taps at mains that should not be overlooked. Take for example a water service pipe which must be run through ground where electricity is escaping under trolley tracks, around power houses, etc. The electricity will enter the pipe and wherever it leaves the pipe a hole is burned. The surface of the pipe in a short time will be full of small pith marks and will soon leak. A good way to add to the life of the pipe under these conditions is to make a star of copper and solder it on to the pipe in the street. Another piece of copper should be put on the pipe near the building. The electricity will leave the pipe by way of the points on the star. This method may not be a cure for electrolysis, but will add to the life of the pipe. Another method employed is to put the pipe in the center of a square box, then fill the box with hot pitch. When this is hardened the pipe will have a covering that will keep out any moisture and bar electricity to a marked degree.

Materials Used.—Galvanized steel pipe does not last [80] under ground.

Galvanized iron, heavy lead, and brass are used. Wooden pipes were once used and stood years of service. No service smaller than 1 1/4 should be used.

When the water service pipe passes through the foundation wall, the pipe should not be built in, but a small arch should be built over the pipe or a piece of XX cast-iron pipe can be used as a sleeve (Fig. 44).

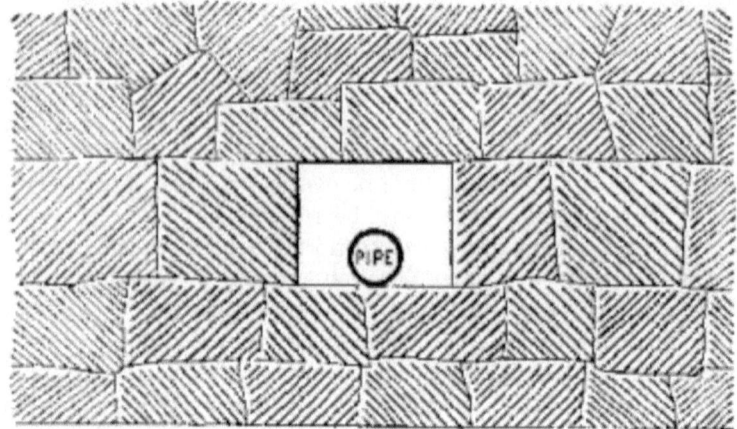

Fig. 44.—Free space around pipe passing through wall.

Points to Remember.—

Sewer Installation

1. *First,* select good sound pipe and fittings.
2. *Second,* locate branch connection in street sewer.
3. *Third,* lay out run of house sewer.
4. *Fourth,* take out necessary permits from departments of sewer.
5. *Fifth,* dig trench in the street, then into the house.
6. *Sixth,* lay pipe and cement joints.
7. *Seventh,* refill trench, tamping every foot.
8. *Eighth,* cast-iron pipe for sewer is found under another heading.

[81]

Water Service

1. *First,* take out necessary permits.
2. *Second,* list material and deliver to job.
3. *Third,* lay out and dig trench.

4. *Fourth*, have main tapped.
5. *Fifth*, lay pipe to curb and test.
6. *Sixth*, fill in street trench.
7. *Seventh*, lay pipe into building and test.
8. *Eighth*, set curb box.
9. *Ninth*, refill trench.
10. *Tenth*, thoroughly consider any special conditions.
11. *Street Sewer.* — Large pipe in streets to receive all soil and waste from buildings.
12. *House Sewer.* — Conveys sewage from building to street sewer, extends from foundation wall to sewer.
13. *Street Main.* — Water pipe running parallel with the street, belonging to the water company.
14. *Service Pipe.* — Runs from the street main into the building.
15. *Corporation Cock.* — Brass stop tapped into street main.
16. *Goose Neck.* — Lead pipe which connects the street main and service pipe.
17. *Trench.* — Hole dug to receive pipe.
18. *Main Tapped.* — Hole drilled through wall of main and a thread made on it while pressure is on.
19. *Curb Cock.* — Brass shut-off placed at curb.
20. *Solder Nipple.* — Piece of brass pipe with thread on one end and plain on the other end which connects lead and iron.
21. *Coupling.* — Fitting which connects two pieces of pipe.
22. *Stop Cock.* — Brass fitting for stopping flow of water.
23. *Curb Box.* — Iron box extending from curb cock to surface.
24. *Curb Key.* — A long key to fit in side of curb box to operate curb cock.
25. *Swab.* — Stick with ball of rags or paper on one end.

CHAPTER VIII

Installing of French or Sub-soil Drains

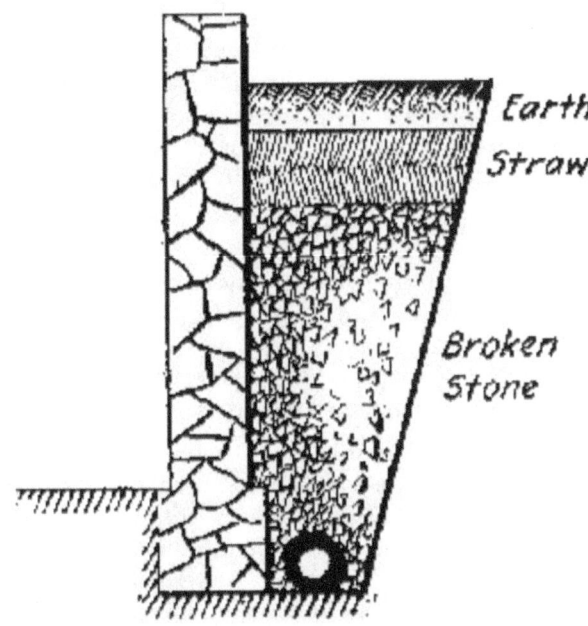

Fig. 45.—Sub-soil drain.

When a building is erected on a site that is wet or springy, some means of carrying off the surplus water in the ground must be provided for, or the basement of the building will be flooded with water. For the thorough understanding of the methods employed in laying a drain of this kind, I will go over it carefully and the beginner can read it and then study it, and understand just how it is done. A site may appear to be dry on the surface of the ground and yet be very wet under the surface. If no information can be had regarding the site, it is always well to drain the site if it is on a slope or near a body of water and on the water shed of a river or lake. If a building is a large one and the foundation goes down very deep, the site should always be drained. The drain is laid under the basement floor and around the outside of the foundation wall on a level with or lower than the basement floor. The value of draining a building site when the building is first started is very often overlooked. The cost of the drain will be saved in a few years as the basement will be free from all excessive dampness. The expense of installing a sub-soil after the building is up and in use is great as well as inconvenient. The drain is called "sub-soil drain" on account of its location under the ground and on account of its duty [83] of taking off all surplus water that is underground. With the surface water taken off by the surface drains and the sub-soil drained by the sub-soil drains, a wet building site can be made practically dry (see Fig. 45).

Materials Used in Sub-soil Construction.—The object of the drain is to collect water and carry it away from the building by means of pipes. Terra-cotta pipes, with or without hubs, are used. Perforated tile pipe is sometimes used. This pipe is unglazed terra-cotta pipe with 1-inch holes in the sides about 3 or 4 inches from the center. These holes allow the surplus water to enter the bore of the pipe and thus be carried off beyond the building site.

When the sub-soil of a small building needs draining, the trenches made for the house drain and its branches are used as a drain in the following manner: The trenches are dug deeper than is required for the house drain. The trenches are then filled to the correct level with broken stones. There is space between these stones for the water to find passage to a point away from the building. When this method is employed, some provision must be made to prevent the house drain from settling. When locating the drain, we must consider approximately the amount of water that is likely to be in the soil and required to be carried off. If there is considerable water, the pipes should extend all around the outside of the building foundation wall, also a main pipe running under the cellar bottom with six branches, three branches on each side.

If there is not a great deal of surplus water in the soil, the drain around the outside of the foundation wall should be put in and one drain line running through the basement will be sufficient.

Laying the Pipe.—The drain pipe should be handled with care, for it is easily broken. The trench should be laid out and dug, then the pipe can be laid in it with a grade toward the outlet or discharge. If pipes with a hub on one end are [84] used, the hub should not be cemented. A little oakum is packed in the hub to steady the pipe and keep sand out, the bottom of joint is cemented, a piece of tar paper can be laid over the top of the joint to keep the sand out. With joints made this way, the water can find its way to the bore of the pipe and yet the sand will be kept out of the pipe. As soon as the water gets into the bore of the pipe it has a clear passageway to some discharge point away from the building. If tile pipes without any hubs are used, some covering should be put around the joint to keep out the sand and still allow the water to find its way into the pipes.

Discharge of Sub-soil Drain.—The water that accumulates in a sub-soil drain must be carried off to some point away from the building. As the pipes are generally under the cellar bottom and under the house drain, it is very evident that this drain cannot discharge into the house drain sewer, directly. If the building site is on a hill, the drain can be carried out and discharged on the surface at a point that is somewhat lower than the level of the pipe under the building. Where this cannot be done, it will be necessary to have the different lines of pipes discharge into a pit. The water is accumulated in this pit until it is filled, then it will automatically empty itself as later explained.

Pit Construction.—The pit for the sub-soil water is constructed of cement. A pit 2 feet square or 2 feet in diameter and 3 feet deep will answer all requirements. A pit of this depth will allow a pitch for all lines of pipe, and is large enough for ordinary installations. The pit is built up to the surface of the cemented floor of the basement and covered with a removable iron cover.

Cellar Drainer or Pump.—A cellar drainer is employed to empty the above-mentioned pit. The cellar drainer works automatically. When the pit is filled with water, the drainer operates and empties the pit and discharges [85] the water into a sink or open sewer connection. When the pit is emptied, the drainer shuts off. The cellar drainer is operated by water pressure. When the valve is opened, a small jet of water is discharged into a larger pipe. The velocity of this small jet of water creates a suction and carries along with it some of the water in the pit. This suction continues until the tank is empty. There should always be a strainer on the suction pipe, also on the supply pipe, to prevent any particles of dirt getting into the valve. The pipes leading to and from the drainer should empty into an open sink where it can be seen. There is a possibility of the drainer valve leaking and then the water pressure will leak through it, causing a waste of water. If this leakage can be seen where it discharges, then the trouble can be rectified. The cellar drainer is connected directly with the water pressure and should have a valve close to the connection to control the supply.

[86]

CHAPTER IX

Storm and Sanitary Drainage with Sewage Disposal in View

The accompanying drawing of storm and sanitary drains should be studied in detail by the reader. The location of each trap and fitting should be studied carefully and the reason that it is put in that particular place should be thoroughly understood. Below, each plan has been taken and gone over in detail, bringing out the reasons for fittings and traps, also the arrangement of the piping.

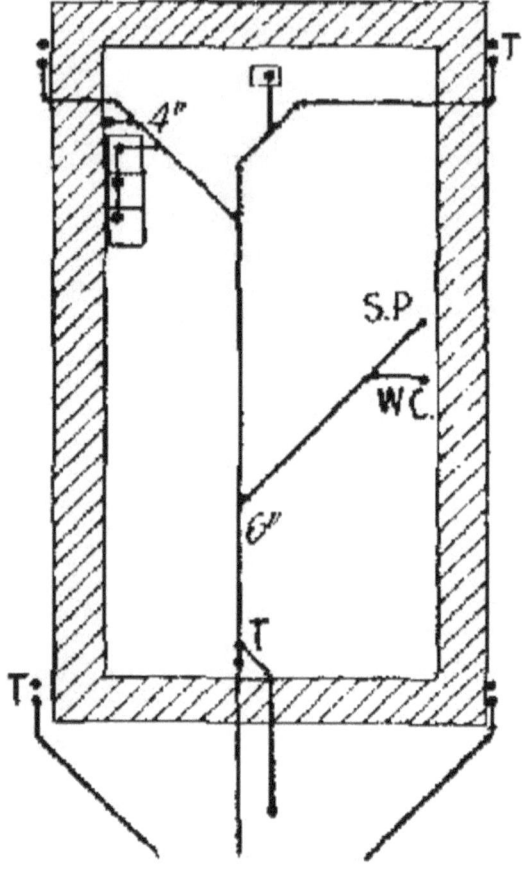

Fig. 46.

The first thing to note in Fig. 46 is the number and kinds of fixtures to be drained. There is in the basement a set of three-part wash trays. This will require a 2-inch waste and a 1 1/2-inch vent. There is in the drawing a 2-inch waste extending to the fixtures above. On the same line is a rain leader with a trap showing also a 4-inch floor drain. There are two 4-inch rain leaders on the opposite corners of the plan, in the rear of the building. There is a 4-inch soil stack for fixtures above and a 4-inch soil stack in the basement on the same line for a basement toilet. On the front there are rain leaders in each corner. These will be connected outside of the house trap (this feature should be noted). The outlets that are to discharge into the house drain are as follows:

- Two 4-inch rain leaders. [87]
- One 2-inch sink waste.
- One 2-inch wash tray waste.
- One 4-inch floor drain.
- One 4-inch soil pipe.
- One 4-inch closet connection.
- Two 4-inch front rain leaders to discharge into house sewer.

If we were to install this job, we would first locate each pipe that enters the house drain. The lowest outlet would be particularly noted, in this case the 4-inch floor drain. From this drain we must make sure that at least 1/4 inch to the foot fall is secured. We must then locate the house sewer where it enters the foundation wall, then the work can be started. I will not attempt to list the material that is necessary for this work, at this time. With all the material at hand the house drain is started. All of this work is installed under the ground, therefore trenches must be dug for all the piping. The plumber must lay these trenches out and in doing so he must have in mind all connections and the fittings he can use so that the trenches can be dug at the right angle. The trenches must be dug allowing a pitch for the pipe. The height of the cellar is 8 feet below the joists. A stick is cut 8 feet long which can be used to get the trenches below the cement floor at the right depth. After the digging is completed, the house trap, which is a 6-inch running trap, is caulked into a length of 6-inch cast-iron pipe. This piece of pipe is pushed out toward the sewer bringing the trap near the foundation wall, on the inside. The fittings and traps and pipe are caulked in place as fast as possible. When possible, the joints are caulked outside of the trench in an upright position. There are a number of different ways to caulk this pipe together, and to make it clear to the beginner just how it is done the following exercise is suggested. This job brings in the caulking of pipes, traps, and fittings in various [88] positions. Two or three can work on this job together. Fig. 47

shows how the pipe and fittings are put together, which needs no further explanation. Therefore, we will go over in detail only the caulking of the joints in the various positions.

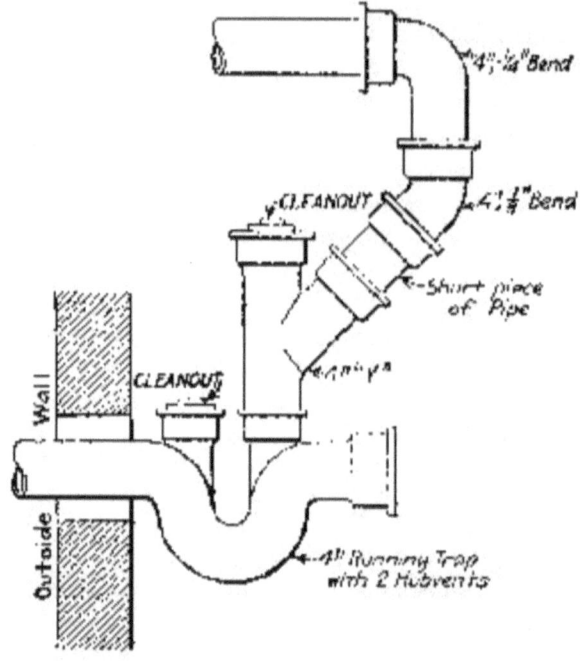

Fig. 47.

Material Needed.—One length of 4-inch extra heavy cast-iron pipe, single hub; two lengths of 4-inch extra heavy cast-iron pipe, double hub; one running trap, one full Y, one 4-inch 1/4 bend; two 4-inch clean-out screws with iron body; one 4-inch vent cap; one 4-inch 1/8 bend; 30 pounds of block lead; 2 pounds of oakum.

Tools Required.—Ladle, asbestos pourer, hammer, cold chisel, yarning iron, two caulking irons, furnace and pot.

The beginner should start at the trap and caulk the joints with the trap held in place. The cold chisel should be sharp as it is used to cut the cast-iron pipe.

To caulk the straight end of cast-iron pipe into the hub [89] end and make a water-tight joint when the pipe is in a vertical position, the spigot end of the pipe is entered into the hub end of another piece. A wad of oakum is taken and forced into the hub with the yarning iron. This piece of oakum is forced to the bottom of the hub, then another piece is put in. The oakum is set and packed by using the yarning iron and hammer. The hub is half filled with oakum. The oakum is forced tight enough to make a water-tight joint. If the oakum used comes in a bale, pieces of it will have to be taken and rolled into long ropes about 18 inches long, the thickness of the rope corresponding with the space between the hub and the pipe. If rope oakum is used, the strands of the rope can be used. After the oakum is well packed into place and the pipe is lined up and made straight, molten lead is poured in and the hub filled. When the lead has cooled, set the lead with the caulking tool and hammer, making one blow on each side of the joint. This sets the lead evenly on every side. If there is any surplus lead, it can now be cut off, using the hammer and cold chisel. The caulking iron is again taken and the lead next to the pipe is tamped, striking the iron with the hammer at an angle to drive the lead against the pipe. After this has been done all around, the caulking iron is held in such a position that the lead around the hub will receive the force of the blow. After this has been done, the center of the lead is caulked and the joint should be tight. With a little practice, this can be done very rapidly. The lead should be poured in while it is very hot. The caulking must not be done by hitting heavy blows as there is a possibility of splitting the hub and thereby rendering the joint unfit for use.

Caulking Joint in Horizontal Position. — It is necessary in a great many cases to caulk a joint in a position where the lead would run out of the joint unless provision were made to hold it in. To caulk a joint in a position of this [90] kind, the pipe is lined up and secured, then the oakum is put in and forced to the bottom of the hub. Then a joint runner, which is an asbestos rope about 2 feet long and about 1 inch in diameter, is fitted around the pipe and forced against the hub where it is clamped by means of an attached clamp. The clamp is put on the top of the pipe and so arranged that a channel will be left in a V shape. This channel allows the hot lead to run between the asbestos runner and the hub. When the lead has had a chance to cool, the asbestos runner is taken off. Where the clamp was, there will be a triangular piece of lead sticking out beyond the face of the hub. This piece has to be cut off, but no attempt should be made to do so until it has been caulked in place and well set; also the rest of the lead should be set. Then the cold chisel can be used and this extra piece of lead taken off. The caulking of the lead in this position is the same as in the previous position and should be carried out closely. The beginner should understand that it is necessary to have not only the joints tight so that running water will not leak out of them, but that the joints must stand a water test. The testing of soil stacks is explained under another heading. The lines of cast-iron pipe depend to a considerable extent upon these joints to make the whole line rigid.

Caulking of Fittings. — The caulking of fittings, while done the same as a straight pipe, is far more difficult. The improper making of these joints is the cause of many leaks. A long sweep fitting is caulked without a great deal of difficulty. If a short bend fitting is used, the matter of caulking is difficult. The fitting is so short that it is almost impossible to get a caulking iron into the throat. The mechanics will have to work at the throat from each side until this part has been sufficiently caulked. I call attention to this point, for I know it to be a failure in a large number of jobs when it comes to put the test on. In [91] order to caulk the fittings, they must be put in their exact location and positions before the lead is poured in, for after the lead is once in the fitting cannot be moved. When there is a series of fittings on a line, their positions in relation to each other must be considered before the lead is poured.

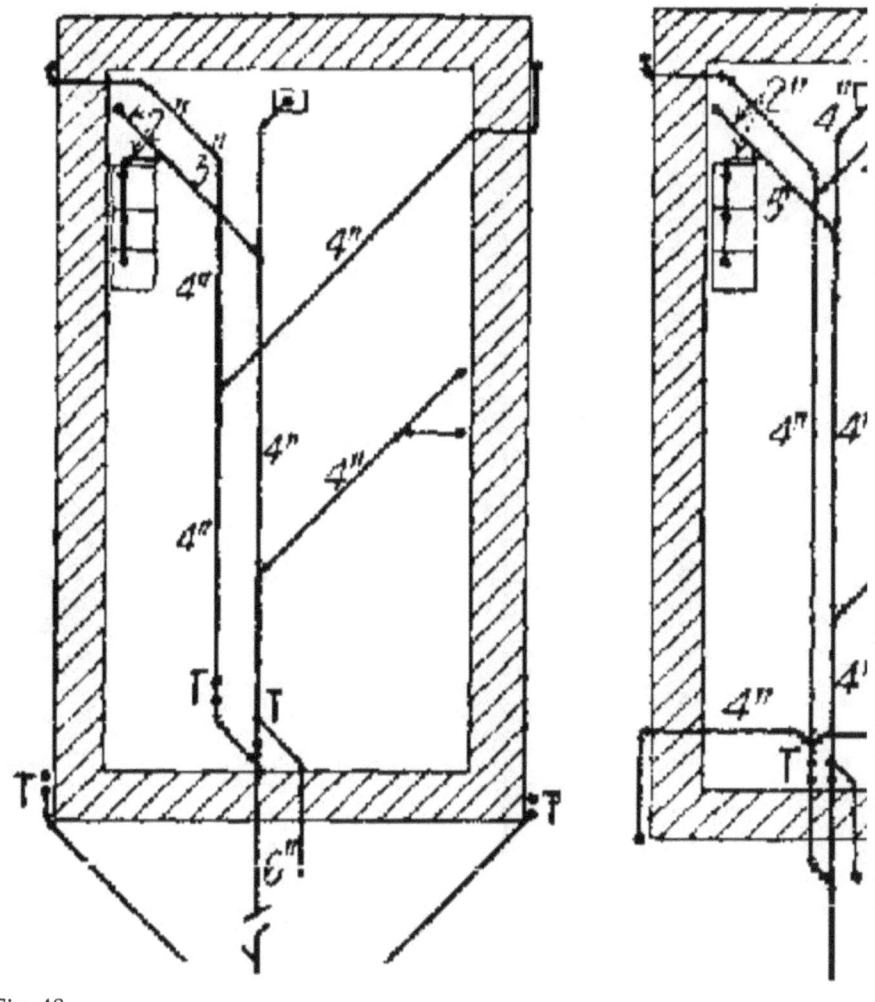

Fig. 48.

Fig. 49.

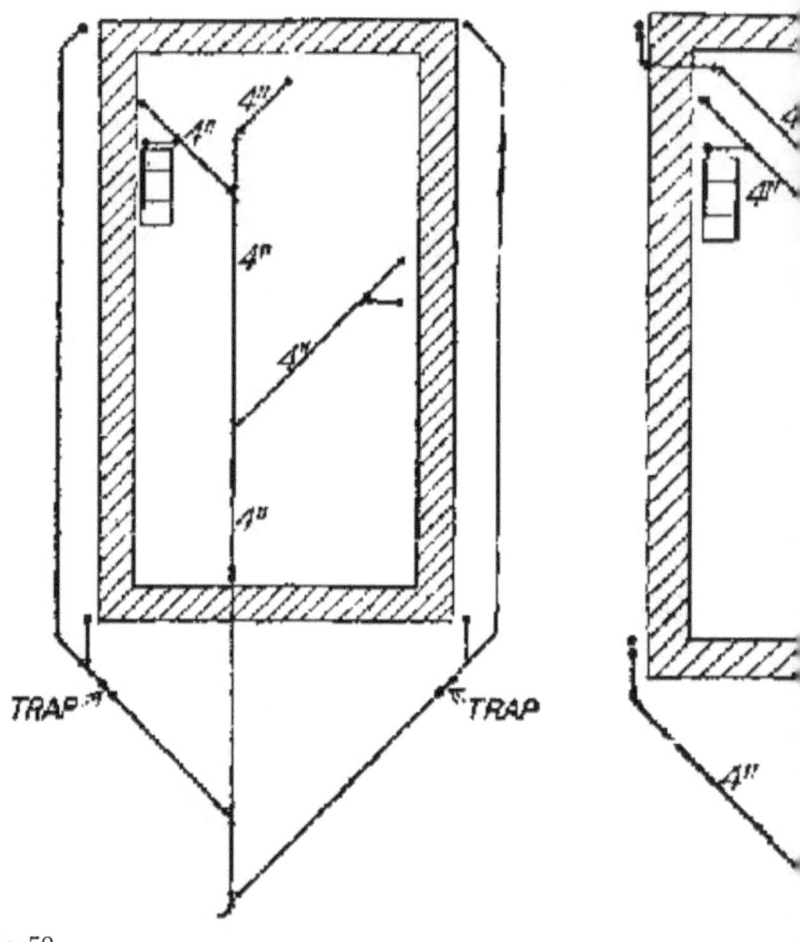

Fig. 50.

Fig. 48 shows the same fixture and stack connections as [92] Fig. 46. Two 4-inch lines run through the cellar, one a sanitary drain, the other a storm drain. Each 4-inch line has an intercepting trap. On the sewer side of these traps the two lines are brought together, beyond which point the two front rain leaders connect; each of the two front leaders is trapped separately.

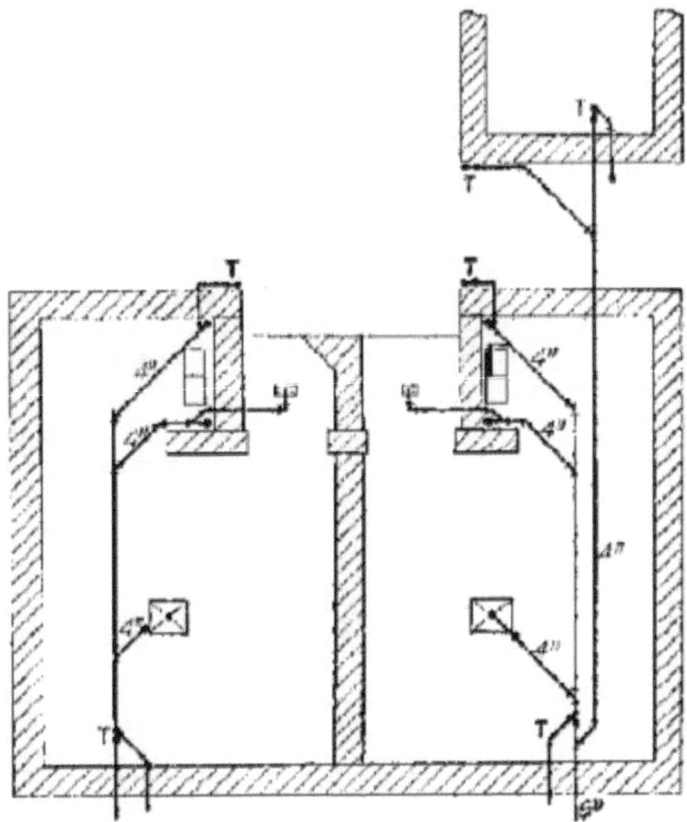

Fig. 52.

Fig. 49 differs from the preceding one in only two points. First, the two front leaders are brought into the cellar and connected into the storm drain on the house side of the intercepting trap. Second, the storm and sanitary drains [93] are connected on the outside of the building.

Fig. 50 shows the same fixtures collected into a 4-inch house drain, and the rain leaders run entirely on the outside of the building. This plan is a good one as all the storm water is kept entirely outside the building. If the storm drains are kept 5 feet away from the cellar walls (see Plumbing Code) the pipes can be of tile. Another good feature of this plan is that all the pipes under the cellar are 4-inch.

Fig. 51 is similar to Fig. 46, the difference being in the location of the floor drain and the connection of the two rear rain leaders, into the house drain.

In Fig. 52 the drains shown take the waste and storm water from the apartment building, also a building set in the rear. The leader pipes in this case are trapped on the outside of the wall. The building in the rear you will note has a separate fresh air inlet and house trap, and the house sewer is continued through the front house and connected into the house drain of the front building, on the sewer side of the intercepting trap.

These drawings should be studied carefully and the student should in each case list correctly all of the material required for the installation of these jobs.

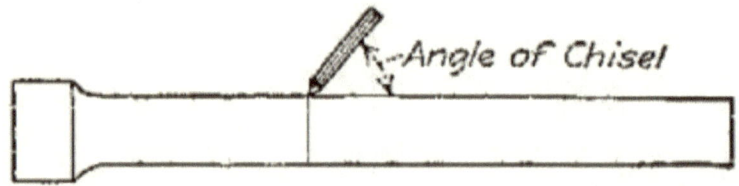

Fig. 53. — Cutting cast-iron pipe.

Cutting Cast-iron Pipe. — To cut cast-iron pipe, a sharp cold chisel and hammer are needed. The pipe is marked all around, just where it is to be cut. Then it is laid with the part of the pipe that is to be cut resting on a block of wood. A groove is cut with the hammer and chisel around the pipe. One person can turn the pipe while the other does the cutting. After a little experience one man can [94] cut and roll the pipe alone. This groove is cut deeper and deeper until the pipe breaks apart. If standard pipe is being cut, a file is generally resorted to for cutting the groove. On account of the lightness of the pipe, a hammer and chisel will crack the pipe lengthwise. When cutting extra heavy cast-iron pipe, a good heavy blow must be struck to cause the chisel to cut into the iron. After a few cuts, the beginner will understand the weight of blow that must be struck to cut the pipe quickly.

[95]

CHAPTER X

Soil and Waste Pipes and Vents. Tests

SOIL PIPES

The term "soil pipes" means pipe that receives the discharge from water closets. The size of a soil pipe for ordinary dwellings should be 4 inches.

Size of Soil Pipes

One to three closets—4-inch XX cast-iron.

Four to eight closets—5-inch XX cast-iron.

Eight to twelve closets—6-inch XX cast-iron.

There are cases when 3-inch XX cast-iron pipe is used, but the practice is not recommended.

The soil pipe should be well supported and held in place. The connection between soil pipe and closet should be of lead to allow for any expansion of settling that might take place.

Material of Soil Pipes.—Soil pipe in common use today is made of light cast iron, tar-coated, extra heavy cast iron uncoated and coated, galvanized wrought-iron pipe, and steel pipe. The best kind to use depends upon the job and place where it is to be used. All kinds of bends and fittings can be had in any of the above-mentioned materials. In choosing the material of the pipe that is best to use, the following points should be carefully considered.

1. *First*, new work or overhauling.
2. *Second*, temporary or permanent job.
3. *Third*, construction of building.
4. *Fourth*, amount allowed for cost of materials on job. [96]
5. *Fifth*, size of job, that is, the number of toilets.
6. *Sixth*, size of chases and pipe partitions.

Location of Soil Pipe.—The location of the soil pipe depends to a great extent upon the location of the toilets. The soil stack should be located on an inside partition. The horizontal pipe should not run over expensively decorated ceilings unless run inside of a trough made of copper or sheet lead. As far as possible, the pipes should be confined, to runs short, and the number of bends reduced.

SOIL-PIPE FITTINGS

Soil-pipe fittings can be had from stock almost to suit the conditions. I will enumerate a few. The names of these fittings should be familiar to the mechanic so that when ordering he can give the correct name. 1/16, 1/8, 1/6, 1/4 bend, sanitary tee, tapped tee, side outlet fitting, return bend, cross branches, double Y, double TY, traps. The uses of these cast-iron fittings perhaps are obvious, but a word about the use of each one will be of service.

The 1/4 bend is used to change the direction of run of pipe 90°. A long-sweep 1/4 bend is used on work requiring the best practice. 1/8, 1/16, and 1/6 bends are used to change the direction of pipe 45°, 22 1/2°, and 16 2/3°. Two 1/8 bends should be used in preference to one 1/4 bend where there is sufficient room. Side outlet 1/4 bend is used for waste connection. They can be had with an outlet on either side of the heel. Their use is not recommended.

Return bends are used on fresh-air inlets. Tees are used for vents only. Ys are used wherever possible. The use of a Y-branch together with an 1/8 bend for a 90° connection with the main line is always preferable to a TY or, as they are commonly called, sanitary T. A tapped fitting gets its name because it is tapped for iron pipe thread. Tapped fittings are used for venting and should not be used for [97] waste unless the tap enters the fitting at an angle of 45°.

These fittings and pipe are joined by first caulking with oakum and pouring, with one continuous pour, the hub full of molten metal. When cool, the lead should be set and then caulked around the pipe and around the hub.

The amount of lead and oakum required for various-sized joints is as follows:

Pipe size............	2	3	4	5	6	8	10	12	15

Pounds of lead....	1½	2¼	3	3 ¾	4 ½	6	7½	9	11¼
Oakum (ounce)...	4	6	8	10	12	16	20	24	30

Rust Joints.—The plumber is called upon to run cast-iron pipe in places where lead and oakum will not be of service for the joints. In cases of this kind, a rust joint is made. This "rust" is made according to the following formula:

- 1 part flour of sulphur.
- 1 part sal-ammoniac.
- 98 parts iron borings (free from grease).

This mixture is made the consistency of cement, using water to mix thoroughly and bring all parts into contact with each other. When it hardens, it becomes very hard and makes a tight joint which overcomes the objections to lead and oakum joints.

WROUGHT-IRON AND STEEL PIPE

This pipe comes in about 18-foot lengths and fittings of the following makes and shapes, and their use is fully explained. The lengths of pipe come with a thread on each end and a coupling screwed on one end. The lengths come in bundles up to 1½-inches and in single lengths over that size. Screw pipe fittings, it will be noted, are called by a different name than cast-iron ones. The fittings in common use today are the 90 degree ell, 45, 22, and 16⅔. The Y and TY, tucker fittings, and inverted Ys are used in [98] practically the same way as the cast-iron fittings. The 90 degree ell, 45, 22, and 16⅔ are used to change the run of pipe that many degrees. All 90 degree fittings, ells, and Ts are tapped to give the pipe a pitch of ¼ inch to the foot. It [99] is better to use two 45 ells to make a 90 bend when it is possible.

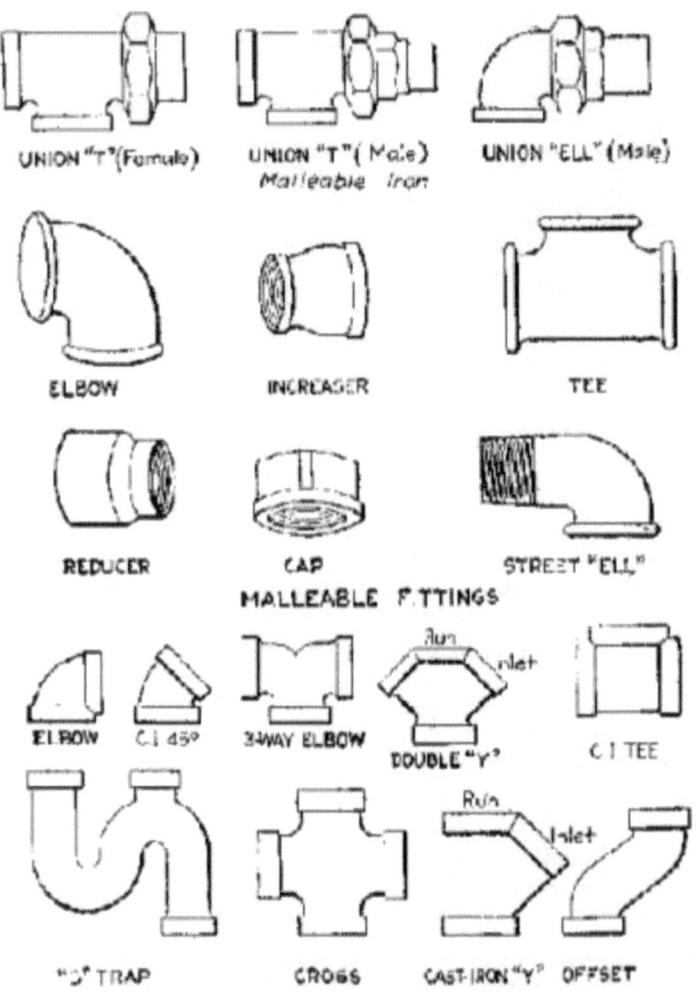

CAST-IRON SCREW FITTINGS Fig. 54.

Inverted Y.—The inverted Y is used in venting to good advantage. The use of these fittings is illustrated in the sketches.

Waste Pipes.—Waste pipes are the pipes that run to or convey the discharge of waste matter to the house drain, from wash trays, baths, lavatories, sinks, and showers.

The usual size of waste pipes is 2 inches. Waste pipes are made of the same material as soil pipe. Lead and brass pipe are also in common use. All exposed waste pipes in bath and toilet rooms are brass, nickel-plated. The waste pipes under kitchen sinks and wash trays are either lead or plain heavy brass. All waste pipes are run with a pitch towards the house trap and should be properly vented as explained under venting. The pipes should be easy of access, with clean-outs in convenient places. The waste pipes under a tile or cement floor should be covered with waterproof paper and a metal V-shaped shield over the entire length. If the waste pipes are over a decorated ceiling they should be in a copper-lined or lead-lined box. This box should have a tell-tale pipe running to the open cellar with the end of the tell-tale pipe left open. If waste pipes are to take the discharge from sinks in which chemicals are thrown, either chemical lead or terra-cotta pipe should be used. If terra-cotta is used, it should have at least 6 inches reinforced concrete around it and the joints of pipe made of keisilgar.

Size of Waste Pipes

Urinals....................	2 inches
Kitchen sink...........	2 inches
Slop sink................	3 inches
Receptacles............	1½ inches
Bath tubs................	1½ inches
Lavatories..............	1½ or 1¼ inches
Wash trays.............	2 inches

Tell-tale Pipe.—The tell-tale pipe is a small pipe that [100] extends from the trough, pan, or box that is under a line of pipe or fixtures to the open cellar. When water is seen running out of this pipe, it shows that a leak exists somewhere in the line of pipe that is in the box or trough. The use of this pipe saves the destruction of walls and ceilings.

VENTS

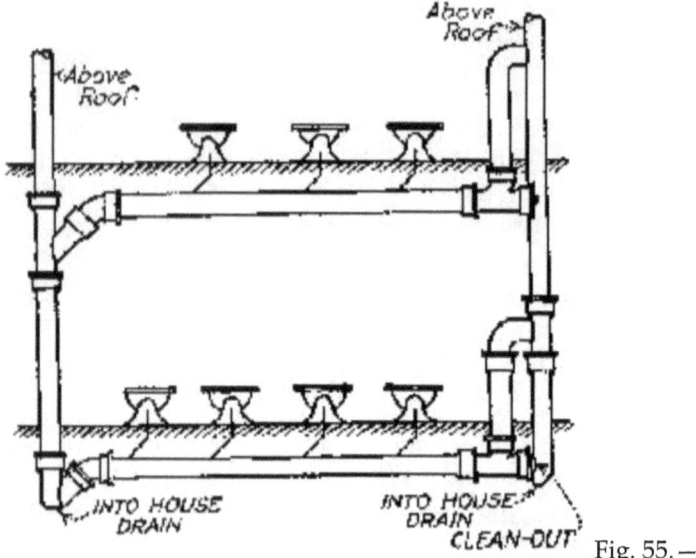

Fig. 55.—Circuit vent.

Vents are the most important pipes in the plumbing system. Modern plumbing successfully attempts to make living in crowded and thickly populated districts, as well as in isolated buildings, free from all unpleasant odors and annoyances. This could not be accomplished without the use of vents. Vents relieve all pressure in the system by furnishing an outlet for the air that is displaced by the waste discharged from the fixtures. Another of its functions is to supply air when syphonic action starts, thereby stopping [101] the action that would break the seal of the trap under fixtures. The pipe extending from top fixture connection, up to and through the roof, is called the ventilation pipe. All vents that do not pass directly through the roof terminate in this ventilation pipe.

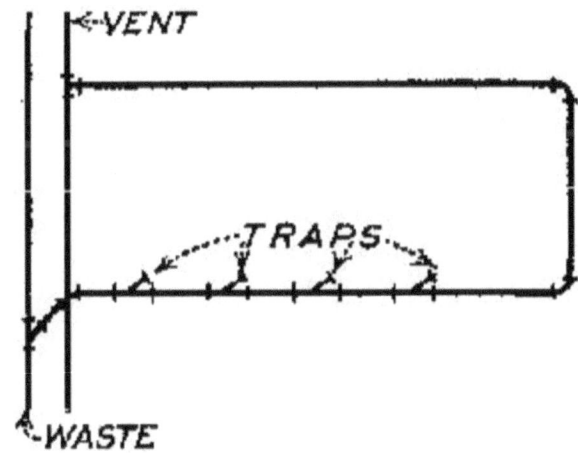

Fig. 56.—Loop vent.

To explain the use of vents, we might well start in the basement of a dwelling house. Suppose there is a set of wash trays in the laundry; the 2-inch trap of these trays should have a 1¼-inch vent pipe leading from the crown of the trap up along side of the stack. On the first floor a 1¼-inch pipe from the crown of the kitchen sink trap will lead into it. Here the pipe should be increased to 2 inches. On the second floor the 1¼-inch pipes leading from the lavatory and bath traps come into it. The vent stack now extends up into the attic and connects with the ventilation pipe. In a general way, the above is an example of venting. The old method of venting was very complicated and is almost beyond describing with the pen.

In common use today, there are several kinds of venting, namely: circuit and loop venting, crown venting, and continuous venting. The *circuit venting*, Fig. 55, is used in connection with the installation of closets. Take a row of toilets in which the waste connection of each closet discharges into a Y-branch, and there will be a series of Y-branches. One end of this series of branches discharges into the main stack while the other end continues and turns up at least to the height of the top of the closet and then enters the main vent stack. When this main vent runs up along side of the main stack and forces the vent pipe connected to the series of Y-branches to travel back, it is [102] called a loop vent. This type of vent supplies air to the complete line of toilets and is very efficient.

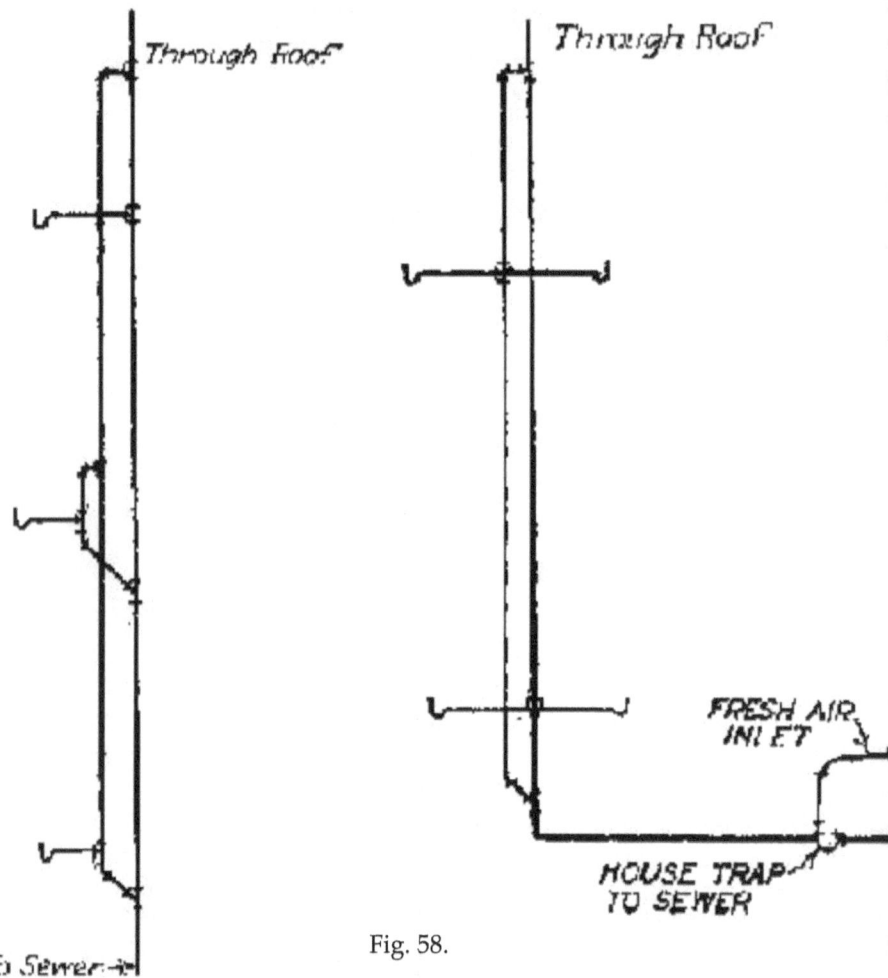

Fig. 58.

Fig. 57.—Continuous vent.

Continuous venting, Figs. 57 and 58, applies more to fixtures other than toilets. A P-trap is used and enters a T in the stack. The lower part of the T acts as and connects with the waste pipe while the upper half is and connects with the vent pipe. A study of the figures will aid the reader to understand thoroughly the above explanations. In continuous venting the waste of the lowest fixture is discharged into the vent pipe and extended to the main waste stack where it is connected. This is done to allow any [103] rust scales that occasionally drop down the vent pipe, and render it unfit to perform its duty, to be washed away into the sewer.

Crown venting, Fig. 59, is as its name implies, a vent that is taken from the crown of the trap, thence into the main vent.

Each one of these methods of venting is used and considered good practice, provided it is properly installed and correctly connected with the use of proper fittings.

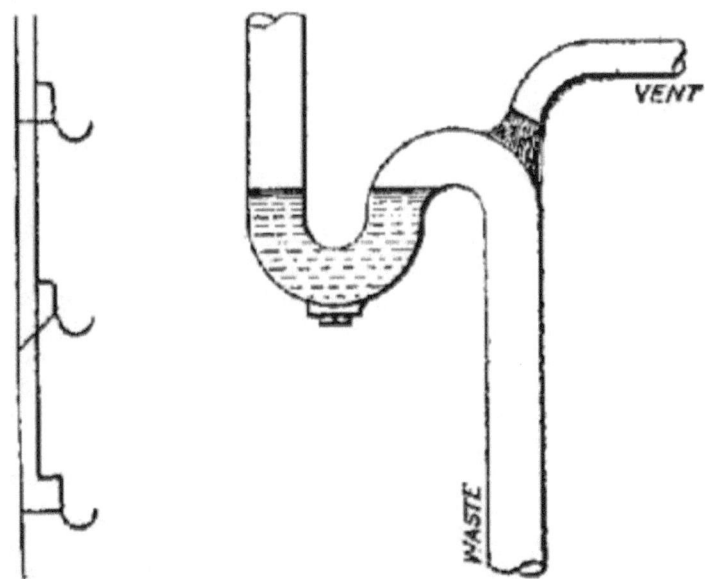

Fig. 59. — Crown venting.

Things to Remember. —

1. *First*, venting is to prevent traps from syphoning.
2. *Second*, also to allow free passage of air.
3. *Third*, circuit vent — loop vent.

4. *Fourth*, continuous venting.
5. *Fifth*, crown venting.
6. *Sixth*, ventilation pipe extends from the top of fixture through roof.

[104]

CHAPTER XI

House Traps, Fresh-air Connections, Drum Traps, and Non-syphoning Traps

The *house trap* is a deep seal trap placed inside the foundation wall, and intersects the house drain and house sewer. The trap is placed at this point for a number of reasons: first, to keep sewer gases from entering the pipes in the house; second, this location is where the house drain ends. This trap should have two clean-outs, one on each side of the seal. The clean-outs should be of extra heavy cast-iron body with a heavy brass screw cap. The cap should have a square nut for a wrench to tighten or unscrew the cap. This cap should be brought up flush with the floor. When a house trap is being set, it is necessary to set it perfectly level, otherwise the seal of the trap is weakened and sewer gases can enter.

Sometimes the trap is located on the house sewer just outside of the foundation wall. In this case, a pit should be built large enough for a workman to get down to it to clean it out when necessary.

A mason's trap was formerly used to a considerable extent, but is very poor practice to use today on modern work. This trap was built square of brick with a center partition. The brick soon became foul and the trap would be better termed a small cesspool than a trap.

Points to Remember about House Traps. —

1. *First*, should be a running trap.
2. *Second*, two clean-outs.
3. *Third*, deep seal, at least 2 inches. [105]
4. *Fourth*, set level.
5. *Fifth*, set inside foundation wall.
6. *Sixth*, accessible at all times.
7. *Seventh*, same size as house drain.
8. *Eighth*, fresh air should connect with it.

FRESH-AIR CONNECTIONS

The term "fresh-air inlet" is, as its name implies, an inlet for fresh air. It is placed directly on the house side of the main trap. The connections made vary considerably. A few good connections in common use are explained below.

When the trap is in place, one of the clean-outs can be used for the fresh air. If this is done, a Y-branch should be placed in the hub of the clean-out. The Y-branch should be used for the fresh air and the run should be used for a clean-out.

A Y-fitting can be inserted directly back of the trap and the branch used for the fresh air. An inverted Y makes a good fitting to use directly back of the trap. These branches should be taken off the top of the pipe. The branch taken off for the fresh-air inlet should not have any waste discharge into it and should not be used for a drain pipe of any description.

The fresh-air inlet should run as directly as possible into the outer air, at least 15 feet from any window. The pipes terminate in a number of different ways, some with a return bend, above the ground, some with a cowl cap, some with a strainer. When necessary to run pipe through the sidewalk, a box of brick is made with a heavy brass strainer fitted level with the sidewalk into which the pipe runs. If the pipe is run into the box on the side a little up from the bottom, the possibility of becoming stopped up or filled up is not great. The fresh-air inlet sometimes terminates above the roof of the building.

Special care should be given this fresh-air inlet as it [106] supplies fresh air to the entire system and thus keeps the pipes in a much better sanitary condition.

Sometimes when the house drain is full of sewage, air is pushed out of the fresh-air inlet and disagreeable odors are evident. This is why it should be located as far as possible from any window. Special care should be taken on the part of the plumber not to locate the fresh-air inlet nearer than 15 feet to the fresh-air intake of the heating system.

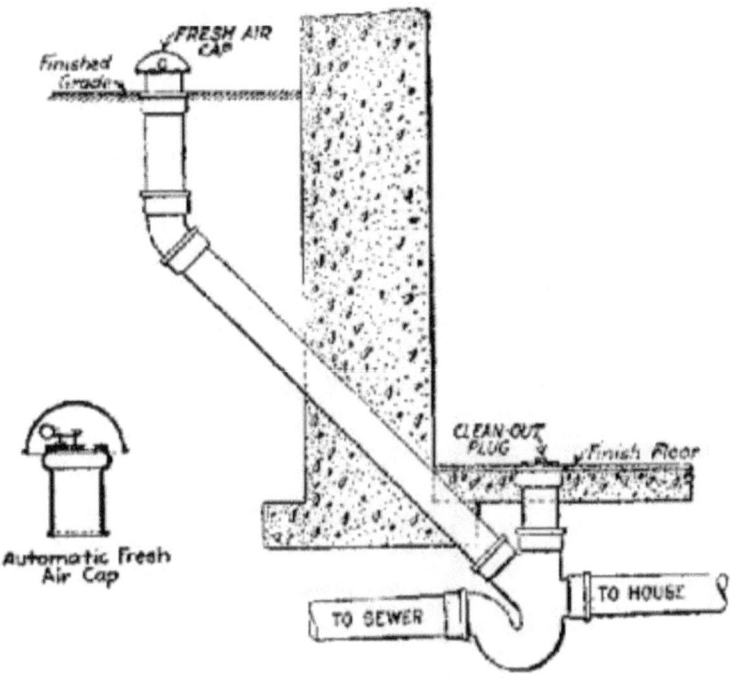

Fig. 60.—Fresh-air inlet.

When the pipe passes through the foundation wall, the same care should be exercised as with other pipes. That is, if the pipe is 4 inches, a sleeve 6 inches should be cut in the wall for the 4-inch pipe to pass through.

Points to Remember about Fresh Air.— [107]

1. *First*, never should be smaller than 4 inches.
2. *Second*, one size smaller than trap.
3. *Third*, location, directly back of trap.
4. *Fourth*, leads to outer air.
5. *Fifth*, keep away from windows and intake of heating system.
6. *Sixth*, always have end of pipe covered with strainer, cowl, or return bend.
7. *Seventh*, make as few bends as possible.

8. *Eighth,* supplies fresh air to system.

DRUM TRAP

The use of the drum trap is very handy to the plumber as well as efficient and practicable when installed. The trap can be purchased without any outlets or inlets, so the plumber can put them in according to the necessary measurements. The making of these traps with lead is explained in the chapter on Wiping Joints. The open end has a brass clean-out screw on it. When this clean-out screw comes below the floor, another brass screw cap and flange is screwed on the floor above the trap so that the clean-out screw in the trap is easily accessible.

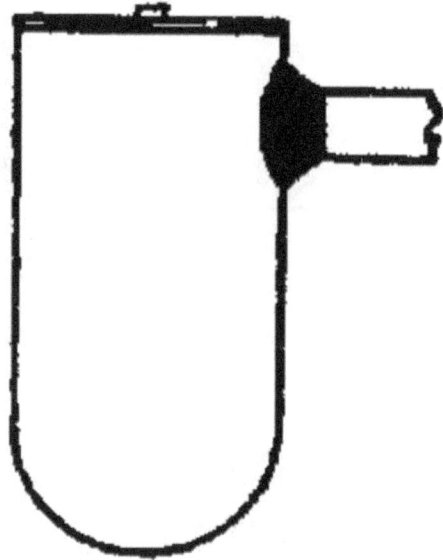

Fig. 61.—Drum trap.

These drum traps are called bath traps as they are used mostly on bath wastes. They should never be installed with the clean-out exposed to the sewer side of the trap. In the best practice, heavy brass drum traps are used.

NON-SYPHONING TRAP

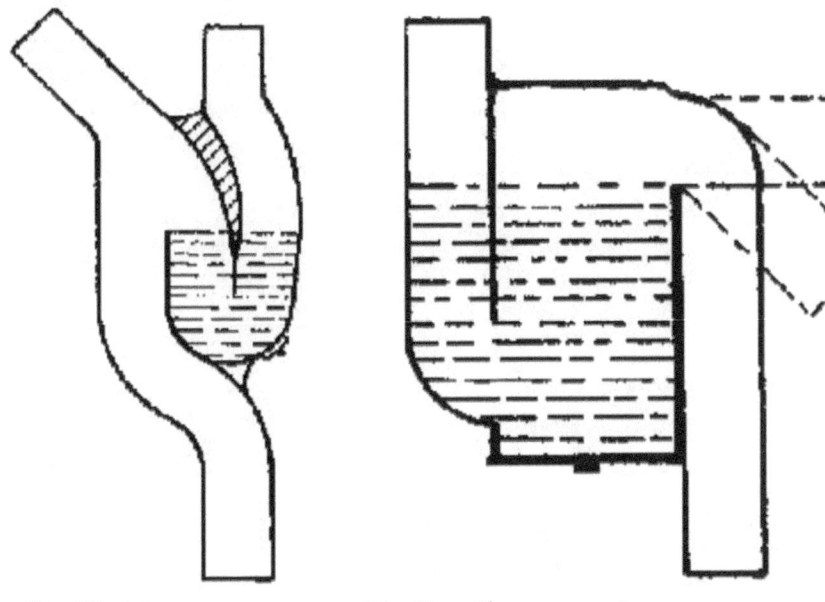

Fig. 62.—Flask trap. Fig. 63.—Clean-sweep trap.

After years of experimenting to produce a trap that would not syphon without venting, we find in use today a large variety of non-syphoning traps. Traps that will hold their [108] seal against all practical forms of syphonic action, or other threatening features, have been made and used and serve the purpose for which they are intended. Various means to prevent the breaking of the seal of these traps are employed. While some depend on a ball or other kind of valve, others rely on partitions and deflections of various kinds. All of these perform the functions for which they are designed, yet the devices employed offer an excellent obstruction for the free passage of waste; therefore, in time, these traps become inoperative. It should be borne in mind that any traps with a mechanical seal or an inside partition are not considered sanitary. The inside partition might wear out or be destroyed and thus break the seal without the knowledge of anyone and allow sewer gas to enter the room. The mechanical device may also be displaced or destroyed, leaving the trap without a seal. If the trap were cleaned out often or examined occasionally, these traps could be used with a greater degree of safety. Some of the forms of non-syphon traps in common use are:

The *Flask Trap*, Fig. 62. This trap gets its name from [109] its shape. There is an inside wall upon which the seal depends. This trap is like the bag trap, only the two inside walls of the pipe are combined into one. This wall should be of heavy cast brass, free from sand holes.

Clean Sweep Trap, Fig. 63. Some clean sweep traps are dependent upon an inside wall for their seals. They are made of 1/2-S, 3/4-S, and full S.

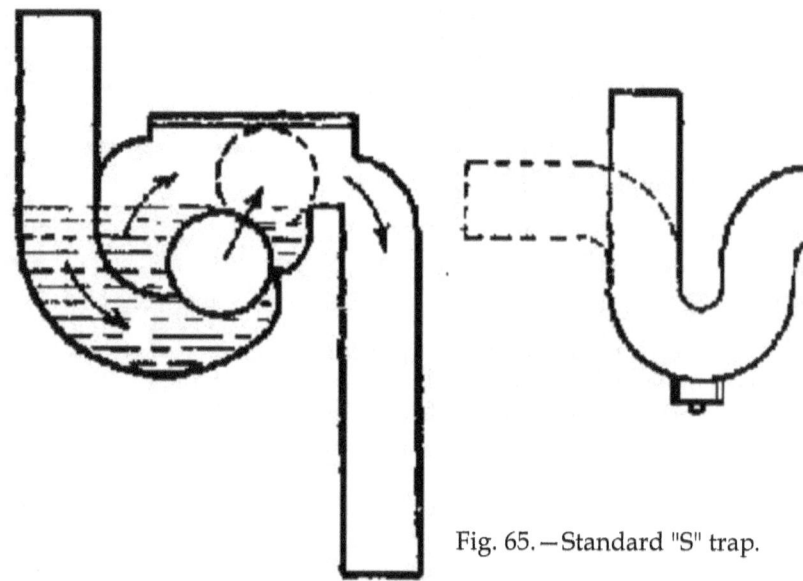

Fig. 65.—Standard "S" trap.

Fig. 64.—Mechanical-seal trap.

Sure Seal Trap. The sure seal trap is designed to be non-syphoning. This trap also has an interior waterway. If this waterway leaks, the trap is unfit for use. If these traps are made as shown in the second sketch with the way inside of a larger pipe, it can be detected if the interior wall leaks.

Centrifugal Trap. The centrifugal trap is made similar to the clean sweep, except that the wall of the inlet pipe is entirely separate from the body of the trap. The inlet enters the body of the trap on a tangent, thus making the trap self-scouring which is a good feature.

[110]

CHAPTER XII

Pipe Threading

The proper cutting of threads on pipe is overlooked by some mechanics. There are many different kinds of dies and different kinds of pipe to contend with. Steel pipe threads very hard and the adjustable dies should be used on it. These dies cut more easily and leave a cleaner thread than other dies when used on steel pipe. When threads are cut on wrought-iron pipe the adjustable dies should be used as they cut a better and cleaner thread than other dies. To preserve the life of the dies and the quality of the thread, oil is used freely while the dies are cutting.

Threads.—The standard thread on pipe and fittings is a right-handed thread. Left threads can be cut on the pipe and the fitting can be tapped with a left thread. When a fitting is tapped with a left thread it is marked so. The following table gives the standard number of threads that a die will or should be allowed to cut on the pipe:

Size	Length, inches	Threads per inch	Threads per end
3/8	9/16	18	10.825
1/2	3/4	14	10.500
3/4	3/4	14	10.500
1	15/16	11 1/2	10.800
1 1/4	1	11 1/2	11.500
1 1/2	1	11 1/2	11.500
2	1 1/8	11 1/2	12.930

To acquaint the beginner with iron pipe work, the following [111] exercise is given. In it there are a great many of the actual problems that come up when the pipe is put in on a job. This is the last exercise that is required in this book. The sketch shows clearly just what the job is and below I have gone over each operation that is necessary to complete the job.

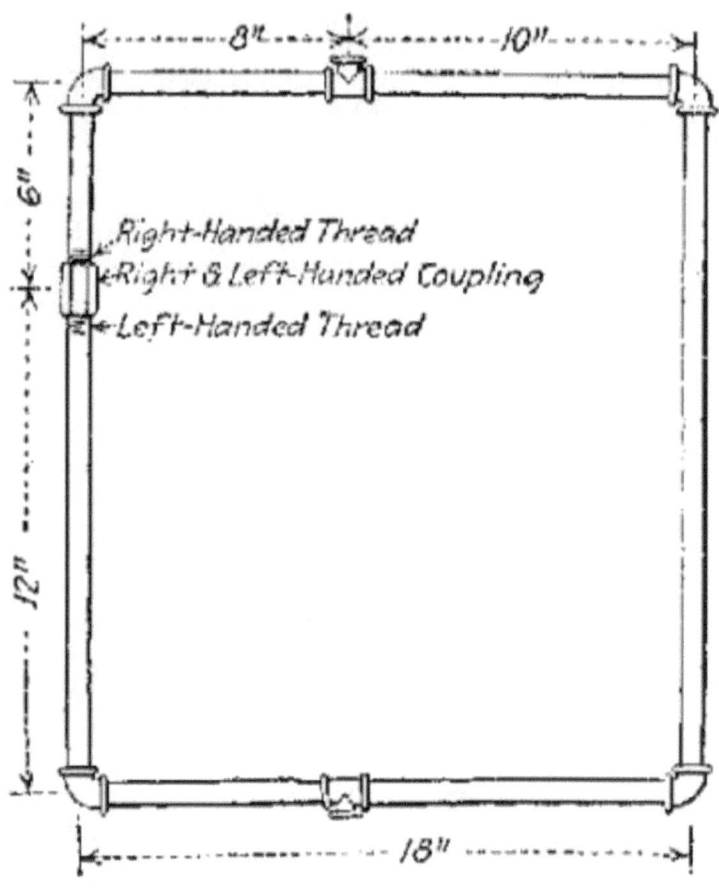

Fig. 67.

Materials Necessary.—Six feet of 1-inch black pipe; four 1-inch black ells; two 1-inch tee; one 1-inch right and left coupling; oil.

Tools Necessary.—Two 14-inch pipe wrenches, vise, pipe cutters, stock and 1-inch follower right and left die and reamer.

The vise is made secure on a bench or post, care being taken before it is put in place to provide room enough to swing the stocks. A length of 1-inch pipe is put into the vise and the vise clamped around it. The end of the pipe that is to be threaded should stick out through the vise [112] about 9 inches. If there is a thread on this end, the dies should be run over it to make sure that it is a standard thread and to clean the threads. Before proceeding further with this exercise the dies and stocks will be described and their use shown.

Dies.—A full set of dies is taken. The full set of stocks and dies is composed of right and left dies from 1/8 inch up to 1 inch, with a guide for each size, also a small wrench with which to turn the set screws. The dies come in sets, two in a set. These are the Armstrong patent that I am describing. Take the stock and the handles, and a set of 1-inch right dies with the guides out of the box. The dies will have marked on them 1" R (if 1-inch left were wanted, the mark would be 1" L). The set screws are taken out of the stock and the dies inserted in their proper place. There is a deep mark on the edge of each die and under it a letter S. This letter means "standard." This mark on the die is set even with a similar mark on the stock and when the set screws are in place and tightened, a standard thread will be cut. There is an adjusting screw on the stock to make the proper adjustment on the dies.

Stock.—The stock is taken and the handles are put into it. There are two sets of set screws on the stock, one set for holding the dies in place and the other set for adjusting the dies. On the stock there is a deep mark to correspond with the standard thread mark on the dies. On the opposite side of the stock there is a place for the follower and a set screw to hold it in place. After the stocks have been looked over and examined thoroughly, the 1-inch right dies are taken and inserted. Then the 1-inch follower is put in place. The tool is now ready to cut a 1-inch thread. Now take a piece of 1-inch pipe at least 15 inches long and put it in the vise, letting it extend out from the vise about 9 inches. The stock is now taken and the follower end is put on the pipe first and the dies brought up in place [113] to cut. The end of the pipe is allowed to enter in between the two dies so that the teeth of each die rest on the pipe. Now, holding the handles of the stock about 6 inches from the body of the stock and standing directly in front of the pipe, push and turn to the right at the same time and the dies will be started. Now put some oil on the dies and turn the stock, taking hold of the ends of the handles and standing at one side. The dies are run up on the pipe until the pipe extends through the face of the dies one thread. Oil is put on the pipe and the dies at least twice during the cutting. When the thread is long enough the stock is turned back a little and then forward a little and the loose chips are blown out from between the dies and pipe. If the dies are set right, a good clean standard thread will have been cut. This thread can now be cut off with the pipe cutters.

Pipe Cutters. — To cut pipe with a one-wheel pipe cutter is a simple matter. I will not dwell at length on the cutter itself. There are one-wheel and three-wheel cutters. We will use a one-wheel cutter tool. This cutter is forced into the surface of the pipe with a set screw having a long tee handle. The pressure that is brought to bear on the pipe while being cut is sufficient to cause a large burr to form on the inside of the pipe. Sometimes the pipe is completely crushed and rendered unfit for use. Therefore the user of these cutters should exercise care when cutting pipe. The pipe is put in the vise and the cutters are so put on the pipe that the pipe will be between the two rollers and the cutter wheel, the cutter resting on the mark that indicates the point at which the pipe is to be cut. The handle is screwed down and the cutters turned around the pipe; each time the cutters make a complete turn the handle is screwed down more. This procedure is continued until the furrow has been cut clear through the pipe.

Cutting and Threading Nipples. — Nipples are short [114] pieces of pipe threaded on each end. Pieces of pipe longer than 6 inches are not called nipples. When a nipple is so short that the threads cut on each end meet in the center of the piece, the nipple is called a "close nipple." When there is a space of about 1/4 inch between the threads, it is called a "space or shoulder" nipple. To cut and thread these nipples a nipple chuck or nipple holder is necessary.

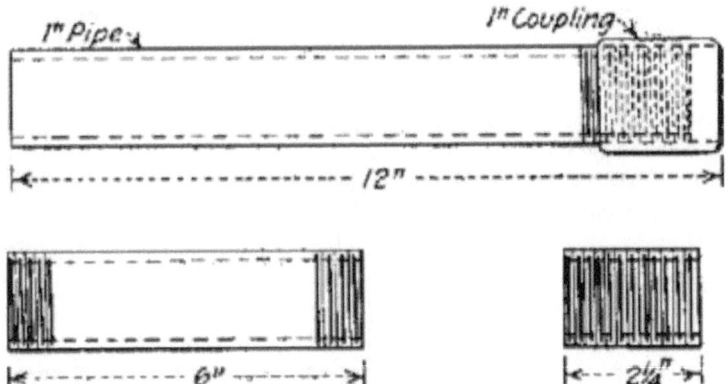

Fig. 68.—Nipple chuck and nipples.

Nipple Holders.—Take a piece of 1-inch pipe about 12 inches long and on one end cut a thread that is 2 inches long. Take a 1-inch coupling and screw it on this end until the end of the pipe is almost through the end of the coupling. At least four threads should be allowed at this end of the coupling. Now we have a piece of pipe 12 inches long having a thread 2 inches long on one end with a coupling on the thread. This is called a nipple holder. Now, to cut a nipple, cut a thread on a piece of pipe and cut the pipe off at any desired length, say 2 inches. Put the nipple holder in the vise with the coupling out from the vise about 8 inches. Take the 2-inch piece of pipe with a thread on one end, screw this thread into the coupling until it touches the pipe that has been screwed through from the other end. Now the stocks having the 1 dies and the follower in are put on the pipe. The follower will not go over the coupling, therefore take the follower out of the stock. Now the [115] stock will slip over the coupling and the thread can be cut. With this procedure a nipple of any length can be cut. There are a number of patented nipple chucks on the market, but as they are not always at hand the above method is resorted to and serves every purpose.

Long Screws.—To cut a long screw which comes in use frequently on vent pipe work, a piece of pipe 12 inches long is taken and a regular length thread is cut on one end, and a thread 4 inches long is cut on the other end. Then a coupling is cut while screwed on a pipe, so that a lock nut about 1/2 inch wide is made. The description and use of these long screws will come under screw pipe venting.

Now that the proper use of the tools has been explained, we will proceed with the exercise according to the sketch. With a length of pipe in the vise and the 1-inch dies in the stock, run over the thread on the pipe. Note that all the measurements are center to center. Screw an elbow on the pipe and measure off the first length, which we will take as 12 inches center to center. Place the rule on the pipe with one end of it at the center of the opening of the elbow just screwed on. Mark 12 inches off on the pipe. This mark represents the center of another ell. Now take another ell and hold the center of one outlet on this mark. It will readily be seen that to have the measurement come right, the pipe must be cut off at a point where it will make up tight when screwed into the ell. Therefore, about 1 inch will have to be cut off, making the pipe 1 inch shorter than where it was first marked. Cut the pipe, and before taking it out of the vise make a thread on the pipe still in the vise. After the thread is cut, take the reamer and ream out the burr that is on the inside of the pipe caused by the pipe cutter. An elbow can be screwed on this pipe. The next measurement is marked off as explained, the pipe cut, then the piece in the vise threaded and reamed. The measurements must be accurate and the dies should be [116] adjusted to cut all threads the same depth. When the measurements are all out, there should be seven pieces of pipe, each piece having one thread. Now the threads on the other end can be cut except the 12 inch piece that screws into the right and left coupling. This thread is a left-handed thread and must be cut with the left dies. Change the dies now to the 1-inch left dies; turn the stock in the opposite direction of the right-hand thread, and the dies will cut the left thread. The pipe and the fittings can easily be put together as shown in the sketch by following the center to center measurements. The right and left coupling is the only fitting that will cause the beginner trouble. A right and left coupling can be used only when there is sufficient *give* to the pipe, that is, the two ends of the pipe to be coupled together are only 1/2 inch apart. To get the coupling in place to start the threads, the pipe must spread apart at least 2 inches. If the pipe cannot be spread that much, a right and left coupling cannot be used. The proper way to make up a right and left coupling is as follows:

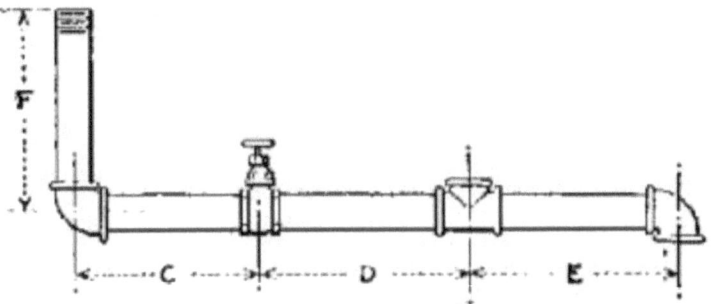

Fig. 69.—F reads center of ell to end, C reads center of ell to center of valve, D reads center of valve to center of T, E reads center of T to center of ell.

Screw home the coupling on the right thread. Mark with a piece of chalk on the coupling and the pipe showing a point on each where the coupling makes tight. Take off the coupling and count the turns and make note of the number. Now do the same on the left thread and make a note of the [117] number of threads. If the left thread has six turns and the right has four and one-half, then to insure that the left thread will be tight when the right thread is, the coupling must be put on the left thread one and one-half turns before it is started on the right thread. Now with four and one-half turns, the right and the left threads will both be tight. A little thought and practice will make this connection clear. If all the measurements in this exercise are not cut accurately, the right and left coupling will not go together.

[118]

CHAPTER XIII

Cold-water Supply. Test

The supplying of cold water to buildings and then piping it to the various fixtures makes a very interesting study. We have gone over the methods of laying and piping for the house service pipe. We will go over the different systems now employed to supply the water, quickly.

Underground Water. — In thinly populated districts the well is still employed to supply water to the building. The water is brought to the surface by means of a large bucket or by means of a pump. A well point can be driven into the ground until water is reached and then the water can be brought to the surface by means of a pump operated by hand or by power. The water can be forced to a tank that is open and elevated, or forced into a tank that is closed and put under pressure. From either tank the water will flow to any desired outlets. A windmill can be employed to furnish power to operate the pump. Water supply that is received directly from underground is by far the best to use. A cesspool or outhouse must not be allowed on the premises with a well, otherwise the well will be contaminated and unfit for domestic use. An open well is not as sanitary as a driven well, as the surface water and leaves, etc., get into it and decay and pollute the water, and soon make it unfit for domestic use.

Streams and Brooks. — The brooks and streams furnish a good source of supply for water to a building or community of buildings. The writer recently worked on a system of piping that supplied 15 or 20 buildings. The water supply came from a brook that was higher than the houses. [119] Each house had a separate pipe leading down from the brook into a tank from which the house was piped. The owner of the brook applied business ethics to the privileges of taking water from it. He had a scale of prices, and the highest-priced location was an inch or so below the bed of the brook, the next price was level with the bottom, the next cheaper 2 inches above the bottom. As the surface was reached, the privilege cost less. In the dry time of the year those at the bottom of the brook always had water while those at the top location had to wait for the water to rise, and had to do without water during the dry time. Where the stream is on a lower level than the building a hydraulic ram can be used.

Rivers and Lakes.—Rivers and lakes make an abundant supply for water systems. A sluggish-moving river is bad, also a river that is used for carrying off the sewage of a town. Special provision is now made for the using of water that is polluted. A lake that is supplied by springs is by far the best source of supply. The water is pumped from the river or lake into a reservoir and then flows by gravity into mains and from the mains into the buildings. The water should always be filtered before it is allowed to enter the distributing mains.

Water Pressure.—Pressure at a fixture or outlet so that the water will flow is generally obtained by the force of gravity. When this method is not sufficient, a pneumatic system is employed. This method is employed to force the water to the top floors or to supply the whole building in high structures. The pneumatic system requires a pump, an air-tight tank, and pipes to the various outlets. The water pumped into the air-tight tank will occupy part of the space generally occupied by the air. The air cannot escape and is, therefore, compressed. Continued pumping will compress the air until the limit of the apparatus is reached. If a valve or faucet that is connected with the [120] tank is opened, the air will expand and force the water out of the opening. This explains in a general way the operation of a pneumatic water-supply system. Water can be pumped into this air-tight tank from a well, cistern, river, lake, or from the city supply mains.

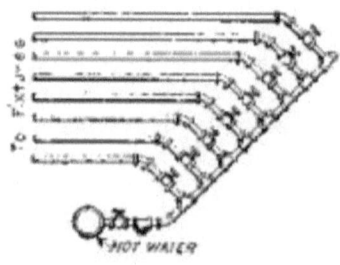
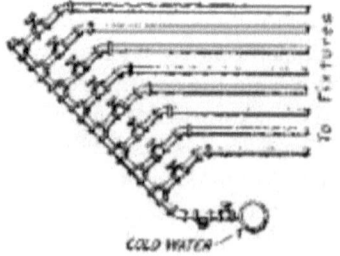

Fig. 70.—"Banjo."

Piping.—From the service pipe on which there has been placed a shut-off, a line of piping, full size, is run through the basement, overhead to a convenient place, perhaps to a partition in the center of the cellar. The pipe is brought down and connected into the end of a header. This header or banjo is made of Ts placed 4 inches center to center. From each T a line of pipe is run to each isolated fixture or set of fixtures (see Fig. 70). A stop and waste cock is placed on each line at such a point that all stop cocks will come in a row near the header. A small pipe is run from the waste of each stop and discharged into a larger pipe which connects with a sink. This way of running pipes while it is expensive makes a very neat and good job. Each stop cock has a tag on it stating explicitly what it controls. If the building is a large one a number of these panelled headers are used. A less expensive way to run this pipe is to branch off from the main at points where the branch pipe will be as short as possible and use as few fittings as possible. Stop and waste cocks are then placed on each [121] branch near the main.

All pipe must follow the direct line of fitting with which it is connected. The line of pipe should be perfectly straight. If it seems necessary to bend the pipe to get around an obstacle, then good judgment has not been used in placing the fitting into which the pipe is screwed. The fitting should be re-located so that the pipe can be run without bending. To have true alignment of pipes the whole job or section of the job must be drawn out on paper first and any obstacles noted and avoided before the piping is cut. This not only saves time but it is also the forerunner of a good job. When getting measurements for piping the same rule or tape should be used to get out the pipe as was used to get the measurements.

The water main and branches that run through the basement of a building are generally hung on the ceiling. Rough hangers of wood, rope, or wire are usually used to hold the pipe in place at first, then neat and strong adjustable hangers are placed every 8 feet apart. There are in use too many kinds of hangers to explain or describe them here. The essential point of all good hangers is to have them strong, neat, and so made that perfect alignment of the pipe can be had. The hangers should be so placed that no strain will come on the fitting or the valves. A hanger should be placed near each side of unions so that when the union is taken apart neither side of the pipe will drop and bend. Hooks and straps should be used to hold vertical pipes rigid and in position. A vertical pipe should be so held in place that its weight will come on the hooks and straps that hold it rather than on the horizontal pipe into which it connects. Where there are six or eight horizontal lines of pipes close together, a separate hanger for each pipe makes a rather cumbersome job and it consumes considerable time to install them properly. A hanger having one support run under all the pipes will allow space for [122] proper alignment and adjustment for drainage. Allowance must be made on all lines of pipe for drainage. When a building is vacant during cold weather, the water is drawn off; therefore, the pipes should have a pitch to certain points where the pipes can be opened and the entire system drained of water.

Kinds of Pipe.—The kind of pipe that is used for cold-water supply depends on and varies according to the kind of water, the kind of earth through which it runs, and the construction of the building. Wrought iron, steel, lead, brass, tin-lined brass, are in use.

The supply pipe to every fixture should have a stop on it directly under the fixture. This will allow the water to be shut off for repairs to the faucet without stopping the supply of other fixtures.

The making of perfect threads on pipe is an important matter, especially on water pipes. If the pipe and the dies were perfect, and the mechanic used sufficient oil in cutting, and the fittings were perfectly tapped to correspond to the dies used on the pipe, of course a perfect union between pipe and fitting would result and the joint would be found to be perfect on screwing the pipe home. As all the above conditions are not found on the job, threads are made tight by the use of red or white lead and oil. The lead is put on the thread and when the thread is made up the lead will have been forced into any imperfection that may be in the threads and the joint will then be water-tight. White lead and oil should be used on nickel-plated pipe as other pipe compounds are too conspicuous and look badly. A pipe compound should be used with discretion, for if too much is put on a burr of it will collect in the bore of the pipe and reduce it considerably. This is not tolerated, so only a small amount is used. Water pipes should be run in accessible places, making it possible to get at them in case of trouble. In climates that have freezing weather water pipes should not [123] be run in outside partitions. If it is found absolutely necessary to do so, as in the case of buildings which have no inside partitions on the first floor, the pipe should be properly covered and protected. The different methods of covering pipes are described in Chapter XV.

[124]

CHAPTER XIV

**Hot-water Heaters. Instantaneous Coil and
Storage Tanks. Return Circulation,
Hot-water Lines and Expansion**

The problem of supplying hot water to plumbing fixtures is one that has required years of study. Each job today demands considerable thought to make it a perfect and satisfactory hot-water system. We will find installations today where the water is red from rust, where there is water pounding and cracking. There are also jobs where the fixtures get practically no hot water. As each job or individual building has its own peculiar conditions, they must be solved by the designer or the mechanic, using the fundamental principles of hot-water circulation. We must first know how much hot water is to be used, also the location of the outlets and the construction of the building; then the size of the pipes and the size of the tanks and their locations can be settled. If the job is a large one, a pump may be employed to insure the proper circulation. After this the pipe sizes and connections can be worked out. The one great enemy of hot-water circulation is air. Therefore, no traps or air pockets should ever appear in the piping system. The boiler, as it is often referred to, is the hot-water storage tank. A copper or iron tank holding sufficient water to supply all fixtures, even when every fixture demands a supply at the same time, is installed in a convenient place and the heating arrangement connected with it. A thermostat can be placed on the system and the temperature of the water controlled. According to the size of the building the problem of furnishing the plumbing [125] fixtures with hot water increases.

Methods of Heating Hot Water.—There are a number of ways of furnishing hot water. Some of the installations are listed below.

A cast-iron or brass water back is placed on the fire pot of a stove or furnace. A separate stove with the fire pot and water jacket is used. A coil of steam pipe is placed inside a hot-water boiler or tank. Gas coil heaters are connected with hot water storage tank, also gas coil instantaneous heaters are connected with the piping direct.

Combinations of the above systems are in use and serve the purpose for which they are intended. For instance, the tank can be connected with a coal range and a gas coil heater, heat being furnished by the range alone or the coil heater alone, or both can be used at the same time. This combination can be connected with the furnace in the cellar, and during the winter months, when the furnace is in use, the water can be heated by the furnace coil. In warm weather, when the furnace is out, the range can supply the necessary heat. In hot weather the coil gas heater can supply the heat.

Connections of Tank and Heating Apparatus.—The ordinary house boiler or hot-water storage tank has four connections, two on top, one on the side, and one on the bottom. The top connections are used for the entrance of cold water into the tank and for the supply of hot water to the fixtures (see Fig. 71). The cold-water inlet has a tube extending into the tank below the side connection. This tube has a small hole filed in it about 6 inches from the top. This hole is to break any syphonic action that may occur at any time. The side connection is for the connection of the pipe coming from the top of the water back. The bottom opening in the tank is for the connection of the pipe coming from the lower water back connection, also for draining the boiler. The circulation of the water can be followed thus: [126] cold water enters the boiler in the tube and discharges into the boiler below the side connection. From here it flows out of the bottom connection into the water back, through the upper connection into the boiler, through the side opening, then to the top of the boiler and out to the fixtures through the fixture supply opening.

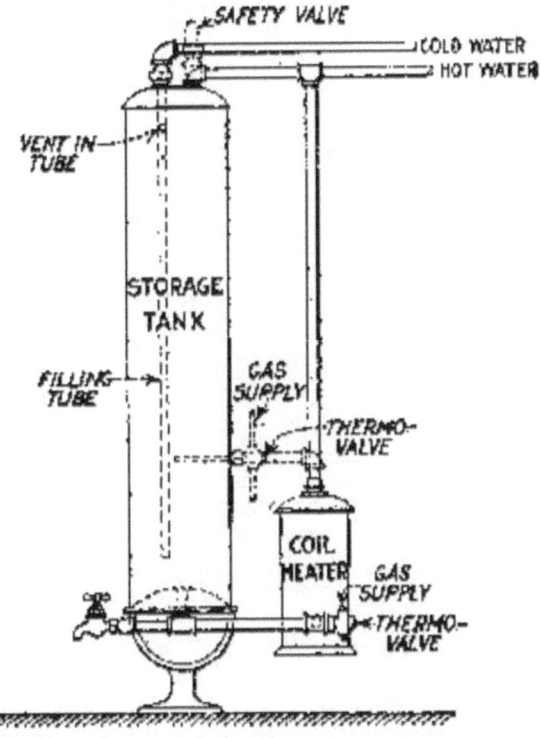

Fig. 71.—
Storage tank, and coil heater with thermostatic control valve.

Fig. 69 shows a thermostatic control valve attached to the bottom of a heater coil, and at the side of storage tank. The best arrangement is at the bottom, for it does not shut off the gas supply until the boiler is full of hot water.

Connecting Tank and Coil Gas Heater.—The boiler and the coil gas heater have a different connection. The bottom of the tank and the bottom of the heater are connected. The top of the heater and the top of the boiler are connected. The accompanying sketch shows how this [127] connection is made. If the tee on the top of the boiler into which the gas-heater connection is made is not the first fitting and placed as close to the outlet as possible, the water will not circulate freely into the boiler. This connection according to the drawing should be studied and memorized.

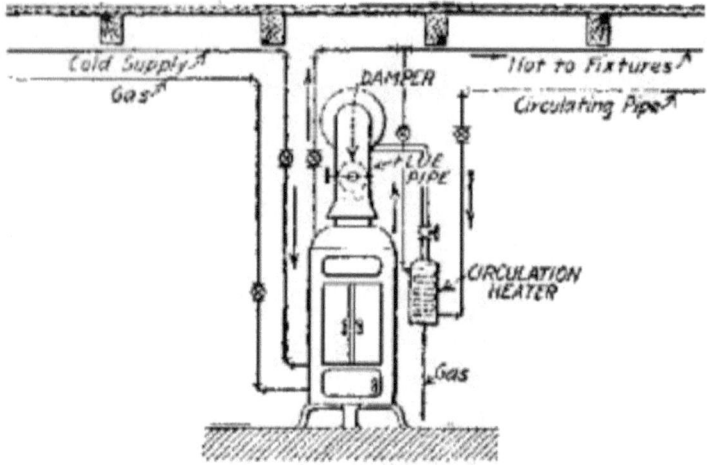

Fig. 72.—Instantaneous gas heater. Showing circulation heater or booster.

Instantaneous Gas-heater Connections.—An instantaneous gas heater is placed in the basement. The copper coil in it is connected at the bottom with the cold-water supply and the top outlet of the coil is connected with the hot-water system of piping. There is no need of a storage tank with this heater. When a faucet is opened in any part of the hot-water piping system, the water passing through the water valve at the heater causes the gas valve to open so that the whole set of burners in the heater is supplied with gas, and the burners are lighted from a pilot light. When the faucet is closed, the gas supply is shut off and the burners are put out. The pilot is lighted all the time. Space will not permit going over these connections in detail. [128] It is a large field and requires considerable thought.

Safety and Check Valves.—When a meter is used on a water system, the water company demands that a check valve be placed on the hot-water system to prevent the hot water from being forced back into the meter in case the pressure got strong enough in the boiler. If a check valve is used for this purpose, or for any other purpose, a safety valve must be placed on the boiler piping system to relieve any excessive pressure that may be caused by having the check valve in use. There is today, with meters of modern type, no reason to use a check valve or a safety valve. If an excessive pressure is obtained in the boiler, it is relieved in the water main.

When water is heated, it expands. If the heat becomes more intense and steam is formed, the expansion is much greater, and some means must be provided to allow for it. This expansion can be allowed to relieve itself in the water main as explained above. When a check valve is placed on the piping, this means of escape is shut off and a safety valve must be employed. Without these reliefs, the pressure would be so great that an explosion would result. When steel pipe and steel boilers are used for storage tanks and connections, the pipe and tank will shortly start to rust and parts of the piping are stopped up with rust scales. The water also becomes red with rust when the water becomes hot enough to circulate. When the pipes are stopped up, steam is formed and a snapping and cracking sound is heard. To avoid these conditions, the piping should be of brass or lead and the storage tank should be of copper. The installation cost of brass and copper is greater than steel, but they will not have to be replaced in two or three years, as is the case with other material. A valve should be placed on the cold-water supply to control the entire hot-water piping system. A pipe with a stop cock should be placed underneath the boiler and should extend [129] into a sink in the basement so that the boiler can be drained at any time for cleaning or repairs.

Connecting with Fixtures.—To have all fixtures properly supplied with hot water it is necessary to run what is termed a circulating pipe. This circulating pipe is a circuit of pipe extending from the top of the boiler to the vicinity of the fixtures and then returning to the boiler and connecting into the pipe leading out of the bottom of the boiler. From this circuit all branches are taken to supply all fixtures requiring hot water. This circulating pipe has hot water circulating through it all the time. Therefore the fixtures are supplied with hot water very quickly. The circulating pipe and its branches are run without any traps or air pockets.

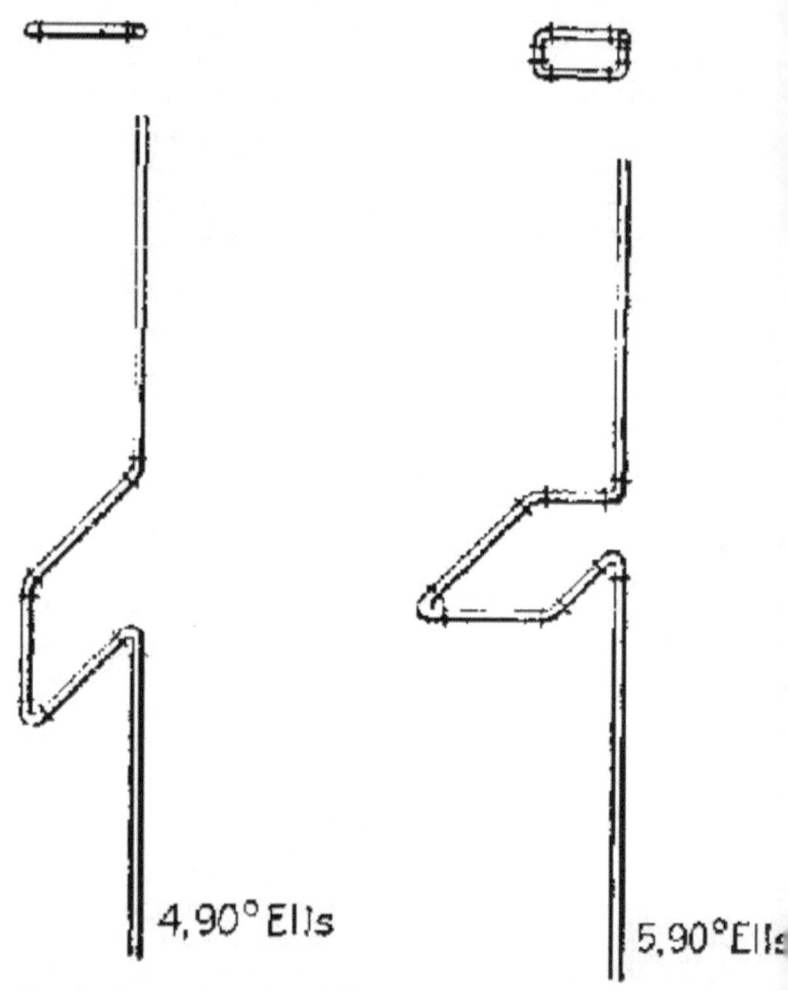

Fig. 73.—Expansion loop. Four 90° ells.

Fig. 74.—Expansion loop. Five 90° e[lls]

When running the piping, it should be borne in mind that not only does the water expand when heated, but the pipe expands also. Therefore due allowance must be made for this expansion. The long risers should have an expansion [130] loop as shown in Figs. 73, 74 and 75. There are installed on some jobs what is known as an expansion joint. This will allow for the expansion and contraction of the pipe. The writer's experience with these joints has not been very satisfactory. After a while these joints begin to leak and they must have attention which in some cases is rather expensive. An expansion loop as shown in the sketch, made with elbows, will prove satisfactory. If the threads on the fittings and pipe are good, no leak will appear on this joint.

All gas heaters must be connected with a flue to carry off the products of combustion.

[131]

CHAPTER XV

Insulation of Piping to Eliminate Conduction, Radiation, Freezing, and Noise

Pipe Covering.—Pipe covering is another important branch of plumbing. A few years ago heating pipes were the only pipes that it was thought necessary to cover. The ever-increasing demands made by the public keep the wideawake plumber continually solving problems. The water running down a waste pipe, for instance, will annoy some people, and provision must be made to avoid this noise or to silence it. This is one of the many problems that the plumber must solve by the use of pipe covering.

Pipes that Need Covering.—First of all, the covering must be put on properly to be of high service. *Hot-water circulating pipes* need covering to reduce the amount of heat loss. If the pipes and the tank are not covered, considerable more fuel will be needed to supply the necessary amount of hot water than if the pipes and tank were covered with a good covering. *Cold-water pipes* need covering in places to keep them from freezing. They also need covering under some conditions to keep them from sweating. They are covered also to prevent the material which surrounds them from coming into direct contact with the pipe. *Waste pipes* need covering to prevent them from freezing and to silence the noise caused by the rush of water through them. *Ice-water pipes* are covered to prevent the water from rising in temperature and to prevent any condensation forming on the pipe. There is need for such a variety of covering that I have listed below some of them and the [132] methods employed for putting them on the pipe.

Magnesia, asbestos air cell, molded asbestos, wool felt, waterproof paper and wool felt, cork, hair felt. These coverings come in the form of pipe covering with a cloth jacket. They also come in the shape of fittings as well as in blocks and rolls of paper, and in powdered form. Any thickness that is desired may be had. The pipe covering is readily put on the pipe. The cloth jacket is pulled back a short distance and the covering will open like a book. It can then be clamped on the pipe and the jacket pulled back and pasted into place. Brass bands, 1 inch wide, come with the pipe covering. These are put on and the pipe covering is then held securely in place. Practically all the coverings are applied in this manner and are made up in 3-foot lengths to fit any size pipe. To cover the fittings and valves, the same kind of sectional covering can be obtained and applied in the same manner as the pipe covering. Plastic covering is often applied to the fittings and molded into the shape of the fitting. The plastic covering comes in bags and is dry. It is mixed with warm water to the consistency of thick cement and applied with a trowel. When the covering is put on the pipes and fittings, it should be done thoroughly to get satisfactory results. Each section of the covering has on one end an extra length of the jacket. This is to allow a lap over on the next section to make a tight joint. If the sections need fitting, a saw can be used and the covering cut to any desired length.

Magnesia covering is employed mostly on steam pipes, especially high-pressure. This material can be had in the shape of pipe covering, in blocks, or cement.

Asbestos air cell covering is employed to cover hot-water circulating pipes. It is constructed of corrugated asbestos paper. This material is manufactured in the sectional pipe covering or in corrugated paper form.

Molded asbestos covering is also used on hot-water pipes, [133] and is manufactured in pipe covering or in blocks.

Wool felt covering is used mostly on hot-water pipes and makes one of the best coverings. It is lined with asbestos paper and covered with a cloth jacket.

Waterproof paper and wool felt is used on cold-water pipes and is made in 3-foot lengths. The covering is lined with waterproof paper and covered with a cloth jacket.

Cork. — A heavy cork covering is one of the best coverings for ice-water pipes, and a light cork covering is used for cold-water pipes. This covering comes in sections as other coverings, also in blocks and sheets.

Hair felt is used to prevent pipes from freezing. It comes in bales containing 150 to 300 square feet of various thicknesses.

[134]

CHAPTER XVI

"Durham" or "Screw Pipe" Work. Pipe and Fittings

"Durham" or "screw pipe" work is the name used to denote that the job is installed by the use of wrought-iron or steel screw pipe. We speak of a "cast-iron job" meaning that cast-iron pipe was used for the piping. A completely different method of work is used when screw pipe is employed for the wastes and vents. When screw pipe is to be used or considered for use, it is well to know something concerning the various makes of screw pipe. Nothing but galvanized pipe is ever used. The value of steel screw pipe and wrought-iron screw pipe should be studied, and every person interested should, if possible, understand how these different pipes are made and how the material of which they are composed is made. In some places one pipe is better than another and a study of their make-up would enlighten the user and allow him to use the best for his peculiar conditions. The maker's name should always be on the pipe. The following table shows the sizes, weights, and thicknesses of screw pipe:

Size	Thickness	No. threads per inch
1 1/4	0.140	11 1/2
1 1/2	0.145	11 1/2
2	0.154	11 1/2
2 1/2	0.204	8
3	0.217	8
3 1/2	0.226	8
4	0.237	8
5	0.259	8
6	0.280	8

Screw pipe work came into common use with the advance [135] of modern steel structures. Some difficulty had been experienced in getting the cast-iron pipe joints tight and to keep the pipe so anchored that it would not crack. The screw pipe was found to answer all of the requirements of modern structures and therefore has been used extensively. The life of screw pipe is not as long as extra heavy cast-iron pipe. This is the only serious objection to screw pipe, which must be renewed after a term of years, while extra heavy cast iron lasts indefinitely. Screw pipe is never used underground. When piping is required underground, extra heavy cast-iron pipe is used.

PIPING

The pipe used in Durham work is galvanized extra heavy, or standard wrought-iron, or steel pipe. It is almost impossible to recognize wrought-iron from steel pipe without the aid of a chemical or a magnifying glass. To test the pipe to distinguish its base, take a sharp file and file through the surface of the pipe that is to be tested. If the pipe is steel, under a magnifying glass the texture of the filed surface will appear to be smooth and have small irregular-shaped grains, and there will also be an appearance of compactness. If the pipe is iron, the texture will have the appearance of being ragged and will show streaks of slag or black. When screw pipe is cut there is always left a large burr on the inside of the pipe. This burr greatly reduces the bore of the pipe and is a source of stoppage in waste pipes. After the pipe is cut this burr should be reamed out thoroughly. One of the strong points of screw pipe is the strength of each joint. Care should therefore be taken to see that perfect threads are cut on the pipe and that the threads of the fittings are perfect. The dies should be set right and not varied on each joint. There should be plenty of oil used when threads are cut so that the thread [136] will be clean and sharp. The follower or guide on stocks should be the same size as the pipe that is being threaded, otherwise a crooked thread will result. If a pipe-threading machine is used, the pipe is set squarely between the jaws of the vise that holds the pipe in place. When cutting a thread on a long length of pipe, the end sticking out from the machine must be supported firmly so that no strain will come on the machine as the pipe turns. It is necessary to cut crooked threads sometimes on the pipe to allow the pipe pitch for drainage or to bring the pipe into alignment where fitting would take up too much room. To cut a crooked thread on a piece of pipe, simply leave the follower out of the stock or put in the size larger. The dies not having a guide will cut a crooked thread. Piping should be run with as few threads as possible. With a thorough knowledge of and the intelligent use of fittings, a minimum number of threads will result.

The pipes in a building are run in compact parallel lines in chases designed especially for them. The tendency is to confine the pipes to certain localities as much as possible. This makes a very neat job and in case repairs are needed, the work and trouble incurred will be confined to one section.

FITTINGS

The fittings used in screw pipe work are cast-iron recess type (see Fig. 54). The fittings are so made that the inside bores of the pipe and the fittings come in direct line with each other, thus making a smooth inside surface at all bends. The fittings are all heavily galvanized. All fittings should be examined on the inside for any lumps of metal of sufficient size to catch solid waste matter, and these must be removed or the fitting discarded. All 90° bends, whether Ts or elbows, are tapped to give the pipe that connects with them a pitch of at least 1/4 inch to the foot. Except where [137] obligatory, 90° fittings should not be used. To make a bend of 90° a Y-branch, a nipple and a 45° bend should be used, or two 45° bends will make a long easy sweep of the drainage pipes and reduce the possibility of stoppage.

Y-branches are inserted every 30 feet at least to allow for a cleanout which can be placed in the branch of the fitting. When a cleanout is placed an iron plug should not be used. These plugs are not removed very often and an iron plug will rust in and be almost impossible to get out. Brass clean-out plugs are used and are easily taken out.

At times it is necessary to connect cast iron and wrought iron, or in a line where a union could be used if the pipe were not a waste pipe, a tucker fitting is used. This fitting is threaded on one end and has a socket on the other to allow for caulking. To get a good idea of all the fittings in general use, the reader should get a catalogue from one of the fitting manufacturers and a survey of it will give the names and sizes of the fittings. However, I show a few common ones. In the writer's opinion, the studying of the catalogue would be of more benefit than a description of fittings at this point. The sizes used and the methods employed to vent the waste-pipe systems are the same as in cast-iron work.

HANGERS AND SUPPORTS

The hanging of screw pipe is a very essential point. The taking of the strain off from a fitting or line of pipe by the use of a hanger is the means of avoiding serious trouble after a job is completed. On horizontal runs hangers are placed not more than 8 feet apart. In a building constructed of wood, the hangers are secured to the joists. In a building constructed of steel beams and concrete the hangers are secured to the steel beams by means of I-beam hangers that clamp on the beams; also in the case of concrete the hangers are extended through the floor and a T [138] is put on the hanger on top of the cement floor; an iron bar or a short piece of smaller pipe run through the T holds the hanger in place and secures it rigidly. The finished floor is laid over the hanger so that it does not show from the top. Hangers on the vertical lines should be placed at every joint and under each fitting. To have the pipe in true alignment, the hangers must be hung and placed in line. Every riser line must have an extra support at the base to avoid any settling of the stack which will crack the fittings and break fixture connections.

MEASUREMENTS

The proper installation of screw pipe work requires getting correct and accurate measurements. Every plumber is or should be able to get correct center to center, center to end, end to end, center to back, and end to back measurements. In Durham work 45° angles are continually occurring. To get these measurements correctly, the following table has been compiled as used by the author and found to be correct. The reader should memorize it so that it may be used without referring to the book.

Measurements

Soil pipe	Screw pipe	Multiplier
1/6 bend	60	1.15
1/8 "	45	1.41
1/12 "	30	2.00
1/16 "	22 1/2	2.61
1/32 "	11 1/4	5.12
1/64 "	5 5/8	10.22

Before any measurements are taken, the lines of pipe are laid out and the position of each fitting known. As I have stated before, the plumber must look ahead with his [139] work. He must have the ability of practically seeing the pipe in place before the work is started. This requires experience and judgment. Before the measurements are taken and the pipe cut consideration must be given to the fact that the fittings and pipes must be screwed into position. Therefore, "can the fitting on the pipe be placed where it is laid out when this is considered?" must be one of the many questions a plumber should ask himself. Allowance must be made for the chain tongs to swing. Whenever possible, a fitting is made up on the pipe while the pipe is in the vise.

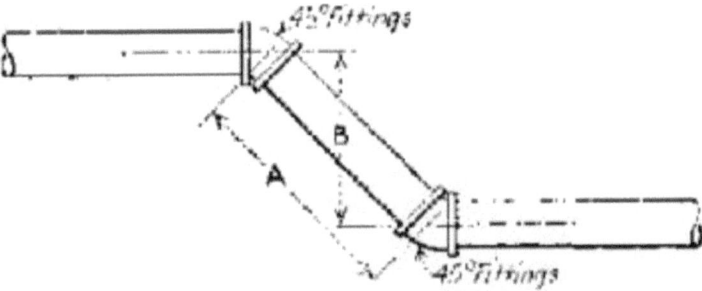

Fig. 76.—The offset is B or 12 inches center to center. The offset is made using 45 degree fittings. Therefore the length of A from the

center of one fitting to the center of the other is B × 1.41 = 12 × 1.41 = 16.92 inches.

FIXTURE CONNECTIONS

The fixture connections when screw pipe is used are necessarily different than when cast-iron pipe is used. A brass nipple is wiped on a piece of lead pipe and then screwed into the fitting left for the closet connection. The lead is flanged over above the floor and the closet set on it. The lead is soldered to a brass flange. The brass flange is secured to the floor and then the closet bowl secured to the brass flange. Another method employed is to screw a brass flange into the fitting so that when it is made up the flange will come level with the floor; the closet bowl is then secured to this flange. There are a number of patented floor flanges for closet bowl connections that can be used to [140] advantage. Slop sinks have practically the same connections as the closets. Other fixtures such as the urinal, lavatory, and bath, can be connected with a short piece of lead wiped on a solder nipple, or the trimmings for the fixture can be had with brass having iron pipe size threads, and the connection can then be made directly with the outlet on the waste line. This is a very general way to describe the connections, but space will not allow a detailed description of these connections. It is always well to allow for short lead connections for fixtures so that the lead will give if the stack settles.

[141]

CHAPTER XVII

Gas Fitting, Pipe and Fittings, Threading, Measuring, and Testing

GAS AND ITS USE IN BUILDINGS

Gas is in common use in all classes of buildings today. Dwellings use it for cooking and illuminating, factories, office buildings, and public buildings for power. In some parts of the country natural gas is found. In these places it is used freely for heating fuel. The actual making of gas is something that every plumber should understand. If space permitted I would describe a gas plant with all of its by-products. However, we shall deal only with the actual installation of gas piping in buildings. Gas mains are run through the streets the same as water mains are run. Branches are taken off these mains and extended into the buildings requiring gas. The gas company generally installs the gas service pipe inside of the basement wall and places a stop cock on it free of charge. This stop that is placed on the pipe is a plug core type, the handle for turning it off is square, and a wrench is required to turn it. The square top has a lug on it. There is also a lug corresponding to it on the body of the valve. When the valve is shut off, these two lugs are together. Each lug has a hole in it large enough for a padlock ring to pass through. This gives the gas company absolute control of the gas in the building.

Setting of the Meter.—Every building that is supplied with gas has a meter that registers the amount of gas consumed. This meter is placed on the service pipe on the [142] house side of the above-mentioned stop cock. This meter is furnished free of charge with a trivial charge made for setting up. The actual setting of this meter is not made until the piping throughout the building has had a thorough and satisfactory test and is found free from all leaks. The meter must be set level on a substantial bracket and in a place, if possible, where it will not require an artificial light to read its dial. The dry meter is usually used in dwellings. The interesting construction and mechanism of this meter cannot be discussed here.

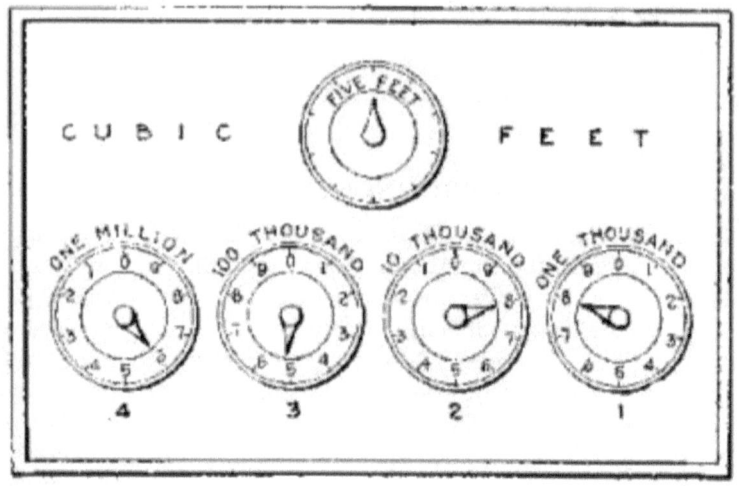

Fig. 77.—Gas-meter dials.

The reading of the dials on a gas meter comes in the province of the plumber and he should be able to read them. The sketch shows the dial plate of a meter. The ordinary house meter has only three recording dials. Large meters have five or more. To read the amount of gas consumed according to the meter we will read the dials as they are indicated on Fig. 77. We will call the four dials No. 1, No. 2, No. 3 and No. 4. In each of these dials a complete revolution of the index hand denotes 1,000, 10,000, 100,000 and 1,000,000, cubic feet respectively. The index hands on No. 1 and No. 3 revolve in the same direction, while No. 2 and [143] No. 4 revolve in the opposite direction. Two ciphers are added to the figures that are indicated on the dials and the statement of the meter will be had. To tell just how much gas has been consumed in a given time, the statement of the meter is taken at the beginning of this given time and at the end of the time. The difference in the figures indicates the number of cubic feet of gas that have been consumed. A gas cock should be placed on the house side of the meter. The dials of meter read 658,800 cubic feet. The dial having the highest number is read first No. 4 dial points to 6, this indicates that No. 3 dial has revolved 6 times. Dial No. 3 reads 5, therefore the reading of dial No. 3 and No. 4 is 65. Dial No. 2 reads 8 making the readings of the three dials 658. Dial

No. 1 reads 8 making the readings of the four dials 6588 add two ciphers to this figure and 658,800 is the correct reading.

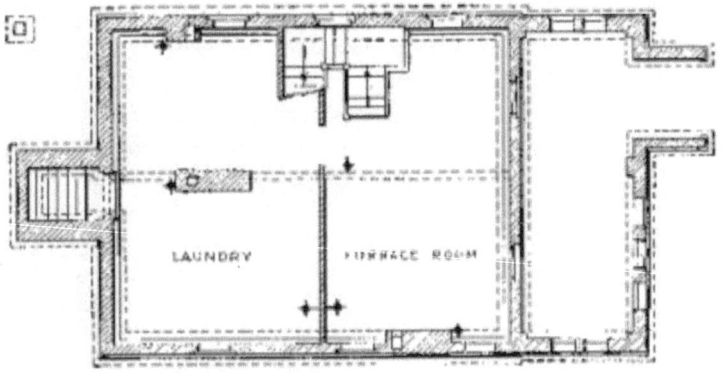

Fig. 78.

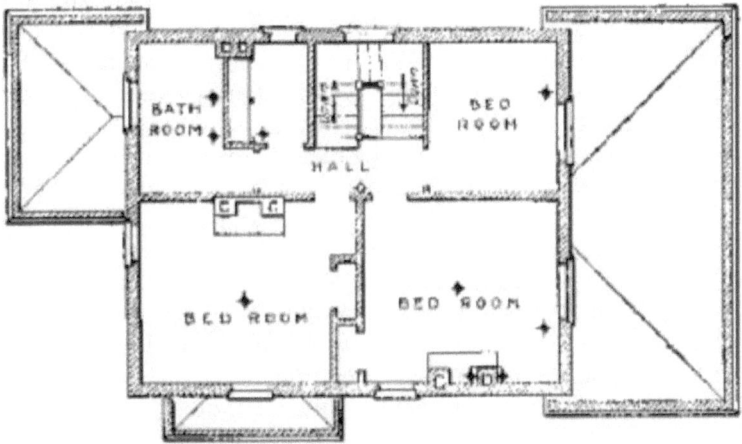

Fig. 79.

Pipe and Fittings.—The pipe used in gas fitting is wrought iron or steel. In special places, rubber hose is used. Brass pipe is occasionally used to advantage. The fittings used in iron pipe gas work should be galvanized. No plain fittings should be allowed. The plain fittings very often have sand holes in them and a leak will result. Sometimes this leak does not appear until after the piping has been in use some time and the expense of replacing the fitting can only be guessed at. By using galvanized fittings, this trouble will be eliminated. All fittings used should be of the beaded type. The fitting and measurement of this work is practically the same as described under iron pipe work. To have the beginner get a clearer idea of gas-piping a building, the piping of the small building sketched will be gone over in detail and studied. One of the first important steps that a gas fitter is confronted with is the locating of the various lights and openings. With these located as shown on the plan, Figs. 78, 79 and 80, we will proceed to work out the piping. The first floor rise will be [144] 1-inch, the second floor will be 1-inch. The horizontal pipe supplying the first floor outlets will be 3/4-inch pipe. The horizontal pipe on the second floor will be 3/4-inch. The balance of the pipe will be 3/8- or 1/2-inch. At this point your attention is called to the sketch of piping, sizes, and measurements. This sketch should be studied and understood in detail. The good mechanic will employ a sketch of this kind when installing any piping. The poor mechanic [145] will take two or three measurements and get them out, put them in, and then get some more. This method is extremely costly and unworkmanlike. There is no reason, except the ability of the workman, why he cannot take a building like the sketch and get all the piping measurements for the job, then get them out, go to the job and put them in. The amount of time saved in this way is so great that a workman should not consider himself a full-fledged mechanic until he can get the measurements this way, and get them accurately. With a tape line, gimlet, and plumb-bob, a mechanic is fully equipped with tools to get his measurements. If the measurements are taken with a tape line, the same tape line should be used when measuring the pipe and cutting it. When laying out the piping, never allow a joist to be cut except within 6 inches of its bearing. It is good policy never to cut timber unless absolutely necessary and then only after consulting with the carpenter. When joists have to be notched they should be cut only on the top side. The pipe as it is put in place should be braced rigidly. Wherever there is an outlet pipe extending through the wall, the pipe should be braced from all sides so that when the fixture is screwed in it will [146] be perfectly rigid.

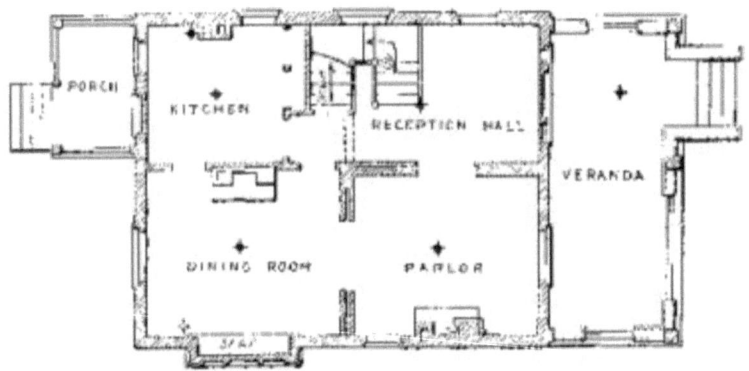

Fig. 80.

The measurements on the piping sketch, Fig. 81, are taken from the accompanying sketch of a dwelling, and if they were to be actually put in, they would fit. The reader would do well to copy this sketch and follow the piping and check the measurements according to the plan, and note how the different risers, drops, etc., are drawn. It is not necessary in a sketch of this kind to draw to a scale. After the different measurements are the letters C.C., E.C., E.E., C.B. and E.B., meaning center to center, end to center, end to end, center to back, and end to back, respectively. Offsetting pipe is a very convenient way of getting the pipe or fittings back to the wall for support. [147] To offset pipe properly and with little trouble, take a piece of scantling 2 by 4 and brace it between the floor and ceiling. Bore a few different-sized holes through it and you will have a very handy device for offsetting pipe. There is a little trick in offsetting pipe that one will have to practice to obtain. The pipe must be held firmly in the place where the pipe is to be bent. Large offsets and bends should not be made; 2 to 4 inches is as large as should be used. Larger offsets that are required should be made with fittings. Always make the offsets true and have the ends perfectly straight. Before putting a piece of pipe permanently in place, always look or blow through it, to ascertain if its bore is obstructed or not. Sometimes dirt or slag will collect and cause stoppage.

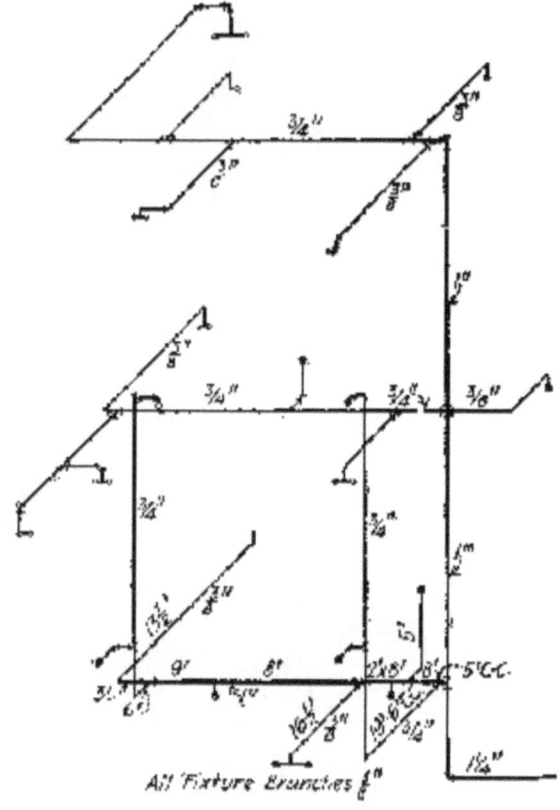

Fig. 81.—Pipe sketch.

Reading the Pipe Sketch.—Vertical lines represent vertical pipes (see Fig. 81). Horizontal lines represent horizontal pipes running parallel to the front. Diagonal lines represent horizontal pipes running from back to front. Any line that is drawn perpendicular to any other line stands for a horizontal pipe. A diagonal line separating a vertical line or horizontal line or set of lines represents a different horizontal plane. With this explanation the sketch will be made clear to one after drawing it. The reader should now take each measurement and check it on the plan. This is easily done by using a scale rule. The height of the ceiling is 8½ feet on the first floor, the second floor is 8 feet. The first floor joists are 10 inches, the second floor joists are 9 inches. An outlet is indicated by a small circle. In the piping sketch, this circle is connected with the riser or drop by a horizontal line. At the junction of these two lines a short perpendicular line is drawn, and indicates the direction of the outlet.

Let me again emphasize the need to understand thoroughly this piping sketch, and to become so familiar with it that it can readily be put to use. The value of a mechanic [148] is determined by the quality and the quantity of work that he can turn out; and a mechanic who can lay out his work and see it completed before he starts, and then proceeds to install his work, is by far of more value to his employer than the man who can see only far enough ahead to cut out two or three measurements and spends most of his time walking between the vise and place of installing the pipe.

Testing.—The system of gas piping must be tested before the pipes have been covered by the advance of building operations. If the job is of considerable size, the job can be tested in sections, and if found tight the sections can be covered. The necessity of having the piping rigidly secured can be appropriately explained here. If the test has been made and the system found tight and some pipe that is not securely anchored is accidentally or otherwise pushed out of place and bent by some of the mechanics working about the building, a leak may be caused and yet not discovered until the final test is made after the plastering is finished. The expense and trouble thus caused is considerable and could have been avoided by simply putting in the proper supports for the pipe.

To test the piping, an air pump and a gage connected with the pipes are placed in a convenient position. The job should now be thoroughly gone over, making sure that all plugs and caps are on and that no outlet is open, also that all pipe that is to be put in has been installed. After this has been attended to, the pump is operated until 10 pounds is registered on the gage. The connection leading to the pump and the piping is now shut off. If the gage drops rapidly, there is a bad leak in the system. This leak should be found without difficulty and repaired. If the gage drops slowly, it denotes a very small leak, such as a sand hole or a bad thread. This kind of leak is more troublesome to find. When it has been found, the pipe or [149] fitting causing the leak should be taken out and replaced. If black caps have been used to cap the outlets, the chances are that a sand hole will be found in one of them. Nothing but galvanized fittings should be used. In case the small leak mentioned above cannot be found by going over the pipe once, there are other means of locating the leak. Two of the methods used, I will explain. If the job is small, each fitting is painted with soap suds until the fitting is found that causes the leak. If the leak is not in the fittings, then the pipe can be gone over in the same way. As soon as the soap suds strikes the leak, a large bubble is made and the leak discovered. It is possible that there are more leaks, so the gage is noted and if it still drops, the search should be continued. The pump should be operated to keep the pressure up to 10 pounds while the search is being made for the leak. When the gage stands at 10 pounds without dropping, the job is then tight. The pump and gage fitting should be gone over first to ascertain if they leak. The other method employed to discover leaks is to force a little ether or oil of peppermint (not essence) into the system by means of the pump. A leak can readily be noted by the odor. To make this method successful, the ether or peppermint should not be handled by the men who are to hunt for the leak. The bottle containing the fluid should not be opened in the building except to pour some into the piping, otherwise the odor will get into the building and as the odor comes out of the leak it will not be noted. For the benefit of the gas fitter, the piping should be tested again after the plastering is completed. The next test is made when the fixtures are put on, and as the piping is tight any leak that develops in this test indicates that the fixtures leak. There are in common use various methods to stop leaks in gas pipe when they are found. If a piece of piping or a fitting is defective, it should be taken out and replaced. This should be remembered so that while the piping is being [150] installed any defects should be noted and the defective fitting or pipe thrown out. Before the gas job is accepted, the gas company will inspect it and look for traps and sags in the pipe. Therefore, the piping should be installed without any traps and it should be arranged to pitch toward the meter, or toward a convenient place from which any condensation can be taken out. If provision is not made for this condensation, it will accumulate and stop the flow of gas.

SHOWER-BATH CONNECTIONS

The sketches show clearly the methods employed to make a shower-bath waste and stall water-tight. The shower bath, as a separate fixture, is in use and the demand for it as a separate fixture is increasing rapidly. This demand comes from the owners of private houses. The plumber must therefore devise some way to make these connections tight and prevent any leak from showing in the room below. This fixture is so constructed that all waste pipes and trap [151] come under the floor level with no way of getting to them from below. Therefore the piping for this fixture must be of a permanent nature. No pipe or trap made of material that is liable to give out in a short time should be allowed under a shower-bath fixture or stall. The two sketches, Figs. 82 and 83 illustrate two methods of connecting and making tight a shower stall. A plumber should always consider it his special duty to make his work complete and free from all objections. He should always prepare for any emergency that may occur in the future. This is rather a big task, yet the plumber when accepting all of his responsibilities has a big task. I state this to the beginner and emphasize the all-important fact that he must learn to perform and think deeply of the elements of plumbing to be able later on to handle successfully the problems that present themselves in the plumbing trade.

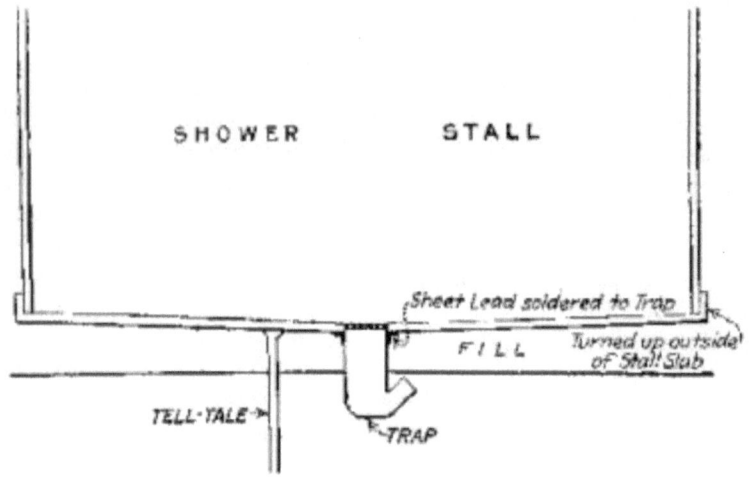

Fig. 82.—Shower stall with lead pan extending outside of stall.

The heavy brass trap shown in the sketch has proved itself very satisfactory and can be made to fit almost any condition of piping or building construction. A flashing of sheet lead is soldered on the trap and carried out to the [152] outside edge of the stall where it is turned up 1 inch, or to the floor level. When the flashing is carried out for only a foot on each side of the trap, the possibilities of a leak are greater.

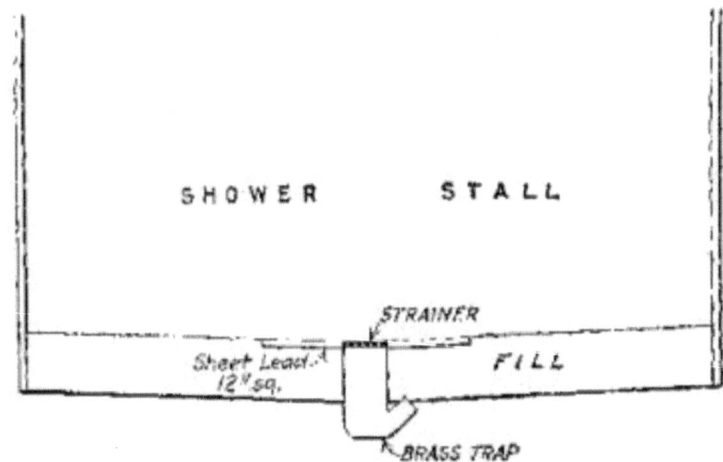

Fig. 83.—Shower stall with lead pan extending six inches beyond strainer.

[153]

CHAPTER XVIII

Plumbing Codes

The work of plumbing has a direct result on the health of the occupants of buildings; therefore in order that the plumbing may not be installed improperly and impair the health of the occupants, it is necessary to provide a code governing the installation of plumbing. Naturally these laws at first were under the control of the health department of cities, but of late years the building departments have assumed control of the codes with the result that coöperation with the building codes is now the practice rather than the exception.

To make certain the carrying out of the plumbing codes, it is required that a plan indicating the run, size, and length of pipes, location and number of fixtures of the prospective job be filed in the building department of the city, before the work is started. If the plan is approved by the plumbing inspector and acceptance is sent, then the work can be started. After a job is completed a test is made and the job is inspected by the plumbing inspector, and if found to meet requirements a written acceptance of the work is given by the building department. An effort is being made throughout the country to have the plumbing codes under State control rather than have a number of different codes in as many different cities and towns. The State code can be so arranged that it will apply to either city or town.

The installation of plumbing varies in different States. In the northern part of the United States all pipes which pass through the roof, if less than 4-inch must be increased to 4-inch. A pipe smaller than 4-inch will be filled with [154] hoar frost during the winter and render the pipe useless to perform its function as a vent pipe. Pipes laid under ground in the Northern States must be at least 4 feet below the surface to protect them from freezing. In the Southern States the frost does not penetrate the ground to such a distance and the pipes can be laid on the surface.

Following is a State or City plumbing code insofar as it relates to the actual installation of plumbing.

Sec. 1. Plans and Specifications.—There shall be a separate plan for each building, public or private, or any addition thereto, or alterations thereof, accompanied by specifications showing the location, size and kind of pipe, traps, closets and fixtures to be used, which plans and specifications shall be filed with the board or bureau of buildings. The said plans and specifications shall be furnished by the architect, plumber or owner, and filed by the plumber. All applications for change in plans must be made in writing.

Sec. 2. Filing Plans and Specifications.—Plumbers before commencing the construction of plumbing work in any building (except in case of repairs, which are here defined to relate to the mending of leaks in soil, vent, or waste pipes, faucets, valves and water-supply pipes, and shall not be construed to admit of the replacing of any fixture, such as water closets, bath tubs, lavatories, sinks, etc., or the respective traps for such fixtures) shall submit to the bureau plans and specifications, legibly drawn in ink, on blanks to be furnished by said board or bureau. Where two or more buildings are located together and on the same street, and the plumbing work is identical in each, one plan will be sufficient. Plans will be approved or rejected within 24 hours after their receipt.

Sec. 3. Material of House Drain and Sewer.—House drains or soil pipes laid beneath floor must be extra heavy cast-iron pipe, with leaded and caulked joints, and carried 5 feet outside cellar wall. All drains and soil pipes connected with main drain where it is above the cellar floor shall be extra heavy cast-iron pipe with leaded joints properly secured or of heavy wrought-iron pipe with screw joints properly secured and carried [155] 5 feet outside cellar wall and all arrangements for soil and waste pipes shall be run as direct as possible. Changes of direction on pipes shall be made with "Y"-branches, both above and below the ground, and where such pipes pass through a new foundation-wall a relieving arch shall be built over it, with a 2-inch space on either side of the pipe.

Sec. 4.—The size of main house drain shall be determined by the total area of the buildings and paved surfaces to be drained, according to the following table, if iron pipe is used. If the pipe is terra-cotta the pipe shall be one size larger than for the same amount of area drainage.

Diameter	Fall 1/4 inch per foot	Fall 1/2 inch per foot
4 inches.....	1,800 square feet drainage	2,500 square feet drainage area
5 inches.....	3,000 square feet drainage	4,500 square feet drainage area
6 inches.....	5,000 square feet drainage	7,500 square feet drainage area
8 inches.....	9,100 square feet drainage	13,600 square feet drainage area
10 inches.....	14,000 square feet drainage	20,000 square feet drainage area

The main house drains may be decreased in diameter beyond the rain-water conductor or surface inlet by permission of the bureau, when the plans show that the conditions are such as to warrant such decrease, but in no case shall the main house drain be less than 4 inches in diameter.

Sec. 5. Main Trap.—An iron running trap with two clean-outs must be placed in the house drain near the front wall of the house, and on the sewer side of all connections. If placed outside the house or below the cellar floor the clean-outs must extend to surface with brass screw cap ferrules caulked in. If outside the house, it must never be placed less than 4 feet below the surface of the ground.

Sec. 6. Fresh-air Inlet.—A fresh-air inlet pipe must be connected with the house drain just inside of the house trap and extended to the outer air, terminating with a return bend, or a vent cap or a grating with an open end 1 foot above grade at the most available point to be determined by the building department.

The fresh-air inlet pipe must be 4 inches in diameter for house drains of 6 inches or less and as much larger as the building department may direct for house drains more than 6 inches in diameter.

Sec. 7. Laying of House Sewers and Drains.—House sewers and house drains must, where possible, be given an even grade to the main sewer of not less than 1/4 inch to the foot. Full-sized "Y"- and "T"-branch fittings for handhole clean-outs must be provided where required on house drain and its branches. No clean-out need be larger than 6 inches.

Sec. 8. Floor Drains.—Floor or other drains will only be permitted when it can be shown to the satisfaction of the department of building that their use is absolutely necessary, and arrangements made to maintain a permanent water seal, and be provided with check or back-water valves.

Sec. 9. Weight and Thickness of Cast-iron Pipe.—All cast-iron pipes must be uncoated excepting all laid under ground, which shall be thoroughly tarred, sound, cylindrical and smooth, free from cracks, sand holes and other defects, and of uniform thickness and of grade known to commerce as extra heavy. Cast-iron pipe including the hub shall weigh not less than the following weights per linear foot:

2-inch pipe...............	5 1/2 pounds per foot.
3-inch pipe...............	9 1/2 pounds per foot.
4-inch pipe...............	13 pounds per foot.
5-inch pipe...............	17 pounds per foot.
6-inch pipe...............	20 pounds per foot.
7-inch pipe...............	27 pounds per foot.
8-inch pipe...............	33 1/2 pounds per foot.
10-inch pipe...............	45 pounds per foot.
12-inch pipe...............	54 pounds per foot.

All cast-iron pipe must be tested to 50 pounds and marked with the maker's name.

All joints in cast-iron pipe must be made with picked oakum and molten lead and caulked gas-tight. Twelve ounces of soft pig lead must be used at each joint for each inch in the diameter of the pipe.

Sec. 10. Wrought-iron and Steel Pipe.—All wrought-iron and steel pipe shall be galvanized. Fittings used for drainage must be galvanized and of recess type known as drainage fittings. [157] All fittings used for venting shall be galvanized and of the style known as steam pattern. No plain black pipe or fittings will be permitted.

Sec. 11. Sub-soil Drains.—Sub-soil drains must be discharged into a sump or receiving tank, the contents of which must be lifted and discharged into the drainage system above the cellar floor by some approved method. Where directly sewer-connected, they must be cut off from the rest of the building and plumbing system by a brass flap valve on the inlet to the catch basin and the trap on the drain from the catch basin must be water-supplied.

Sec. 12. Yard and Area Drains.—All yard, area and court drains when sewer-connected must have connection not less than 4 inches in diameter. They should be controlled by one trap—the leader trap if possible. All yards, areas and courts must be drained. Tenement houses and lodging houses must have yards, areas and courts drained into sewer.

Sec. 13. Use of Old Drains and Sewers.—Old house drains and sewers may be used in connection with new buildings or new plumbing, only when they are found, on examination by the department of building, to conform in all respects to the requirements governing new sewers and drains. All extensions to old house drains must be of extra heavy cast-iron pipe.

Sec. 14. Leader Pipes.—All building shall be provided with proper metallic leaders for conducting water from the roofs in such manner as shall protect the walls and foundations of such buildings from injury. In no case shall the water from such leaders be allowed to flow upon the sidewalk but the same shall be conducted by a pipe or pipes to the sewer. If there is no sewer in the street upon which such building fronts, then the water from said leader shall be conducted, by proper pipes below the surface of the sidewalk, to the street gutter.

Inside leaders shall be constructed of cast iron, wrought iron or steel, with roof connections made gas-and water-tight by means of heavy copper drawn tubing slipped into the pipe. The tubing must slip at least 7 inches into the pipe. Outside leaders may be of sheet metal, but they must connect with the house drain by means of cast-iron pipe extending vertically 5 feet above grade [158] level, where the building is located along public driveways or sidewalks. Where the building is located off building line, and not liable to be damaged the connection shall be made with iron pipe extending 1 foot above the grade level.

All leaders must be trapped with running traps of cast iron, so placed as to prevent freezing.

Rain leaders must not be used as soil, waste or vent pipes, nor shall such pipes be used as rain leaders.

Sec. 15.—Exhaust from Steam Pipes, Etc.—No steam discharge or exhaust, blow-off or drip pipe shall connect with the sewer or the house drain, leader, soil pipe, waste or vent pipe. Such pipes shall discharge into a tank or condenser, from which suitable outlet to the sewer shall be made. Such condenser shall be supplied with water, to help condensation and help protect the sewer, and shall also be supplied with relief vent to carry off dry steam.

Sec. 16. Diameter of Soil Pipe.—The smallest diameter of soil pipe permitted to be used shall be 4 inches. The size of soil pipes must not be less than those set forth in the following tables.

Maximum number of fixtures connected to:

Size of pipe	Waste and soil combined		Soil pipe alone	
	Branch fixtures	Main fixtures	Branch water closets	Main water closets
4-inch	48	96	8	16
4.5-inch	96	192	16	32
6-inch	268	336	34	68

If the building is six (6) and less than twelve (12) stories in height, the diameter shall not be less than 5 inches. If more than twelve (12) it shall be 6 inches, in diameter. A building six (6) or more stories in height, with fixtures located below the sixth floor, soil pipe 4 inches in diameter will be allowed to extend through the roof provided the number of fixtures does not exceed [159] the number given in the table. All soil pipes must extend at least 2 feet above the highest window, and must not be reduced in size. Traps will not be permitted on main, vertical, soil or waste-pipe lines. Each house must have a separate line of soil and vent pipes. No soil or waste line shall be constructed on the outside of a building.

Fixtures with:

- 1-1¼-inch traps count as one fixture.
- 1-1½ " traps count as one fixture.
- 1-2" traps count as two fixtures.
- 1-2½ " traps count as three fixtures.
- 1-3" traps (water closets) count as four fixtures.
- 1-4" traps count as five fixtures.

Sec. 17. Change in Direction.—All sewer, soil, and waste pipes must be as direct as possible. Changes in direction must be made with "Y"- or half "Y"-branches or one-eighth bends. Offsets in soil or waste pipes will not be permitted when they can be avoided, nor, in any case unless suitable provision is made to prevent the accumulation of rust or other obstruction. Offsets must be made with fourth degree bends or similar fittings. The use of T "Y"s (sanitary Ts) will be permitted on upright lines only.

Sec. 18. Joints on Soil and Waste Pipes.—Connection on lead and cast-iron pipe shall be made with brass sleeve or ferrule, of the same size as the lead pipe inserted in the hub of the iron pipe, and caulked with lead. The lead must be attached to the ferrule by means of a wiped joint. Joints between lead and wrought-iron pipes must be made with brass nipple, of same size as lead pipe. The lead pipe must be attached to the brass nipple by means of a wiped joint. All connections of lead waste pipes must be made by means of wiped joints.

Short nipples on wrought-iron and steel pipes must be of thickness and weight known as "extra heavy" or "extra strong."

Brass ferrules must be best quality, extra heavy cast brass, not less than 4 inches long and 2 1/4, 3 1/2 and 4 1/2 inches in diameter and not less than the following weights:

[160]

Diameters	Weights
2 1/4 inches	1 pound 0 ounce.
3 1/2 inches	1 pound 12 ounces.
4 1/2 inches	2 pounds 8 ounces.

Sec. 19. Solder Nipples.—Solder nipples must be heavy cast brass or of brass pipe, iron pipe size. When cast they must be not less than the following weights:

Diameters	Weights
1 1/2 inches	0 pound 8 ounces.
2 inches	0 pound 14 ounces.
2 1/2 inches	1 pound 6 ounces.
3 inches	2 pounds 0 ounce.
4 inches	3 pounds 8 ounces.

Sec. 20. Brass Clean-outs.—Brass screw caps for clean-outs must be extra heavy, not less than 1/8 inch thick. The screw cap must have a solid square or hexagonal nut not less than 1 inch high and a least diameter of 1 1/2 inches. The body of the clean-out ferrule must be at least equal in weight and thickness to the caulking ferrule for the same size pipe.

Sec. 21. Lead Waste Pipe.—All lead waste, soil vent and flush pipes must be of the best quality, known in commerce as "D," and of not less than the following weights per linear foot:

Diameters	Weights
1 1/4 inches	2 1/2 pounds.
1 1/2 inches	3 pounds.
2 inches	4 pounds.
3 inches	6 pounds.
4 inches	8 pounds.

All lead traps and bends must be of the same weight and thicknesses as their corresponding pipe branches.

Sec. 22. Roof Flashers.—Sheet lead for roof flashings must be 6-pound lead and must extend not less than 6 inches from the pipe and the joint made water-tight.

Sec. 23. Traps for Bath Tubs, Water Closets, Etc.—Every sink, bath tub, basin, water closet, slop hopper, or fixtures having a waste pipe, must be furnished with a trap, which shall be placed as close as practicable to the fixture that it serves and in no case shall it be more than 1 foot. The waste pipe from the bath tub or other fixtures must not be connected with a water-closet [161] trap.

Sec. 24. Size of Horizontal and Vertical Waste Pipes, Traps and Branches.—

Horizontal and vertical	Number of small fixtures
1¼-inch	1
1½-inch	2
2 -inch	3 to 8
2½-inch	9 to 20
3 -inch	21 to 44

If building is ten (10) or more stories in height, the vertical waste pipe shall not be less than 3 inches in diameter. The use of wrought-iron pipe for waste pipe 2 inches or less in diameter is prohibited.

The size of traps and waste branches, for a given fixture, shall be as follows:

Kind of fixtures	Size in inches	
	Trap	Branch
Water closet	3	4
Slop sink with trap combined	3	3
Slop sink ordinary	2	2
Pedestal urinal	3	3

Floor drain or wash..................................	4	4
Yard drain or catch basin........................	4	4
Urinal trough..	2	2
Laundry trays, two or five......................	2	2
Combination sink and tray (for each fixture).....	1½	2
Kitchen sinks, small.................................	1½	1½
Kitchen sinks, large hotel, etc..................		
Kitchen sinks, grease trap........................		2
Pantry sinks..	1½	1½
Wash basin, one only...............................	1¼	1¼
Bath tub..	2	2
Shower baths..	1½	1½
Shower baths, floor..................................	2	2
Sitz bath..	1½	1½
Drinking fountains...................................	1¼	1¼

Sec. 25. Overflow Pipes.—Overflow pipes from fixtures must [162] in all cases be connected on the inlet side of the traps.

Sec. 26. Setting of Traps Without Re-vent.—All traps must be substantially supported and set true with respect to their water levels. No pot, bottle or "D" trap will be permitted nor any form of trap that is not self-cleaning, nor that has interior chambers or mechanism nor any trap except earthenware ones that depend upon interior partitions for a seal. In case there is an additional fixture required in building and it is impossible to re-vent pipe for the trap, the building department may designate the kind of trap to be used. This shall not be construed to allow traps without re-vents in new buildings.

Sec. 27. Safe and Refrigerator Pipes.—Safe-waste pipes must not connect directly with any part of the plumbing system. Safe-waste pipes must discharge over an open, water-supplied, publicly-placed, ordinary-used sink, placed not more than 3 1/2 feet above the cellar floor. The safe waste from a refrigerator must be trapped at the bottom of the line only and must not discharge upon the ground floor, but over an ordinary open pan, or some properly-trapped, water-supplied sink, as above. In no case shall the refrigerator waste pipe discharge into a sink located in a living room.

The branches on vertical lines must be made by means of "Y" fittings and be carried to the safe with as much pitch as possible. Where there is an offset on the refrigerator waste pipe in the cellar, there must be clean-outs placed. These clean-outs must be of brass.

In tenement and lodging houses the refrigerator waste pipe must extend above the roof, and not be larger than 1 1/2 inches and the branches not smaller than 1 1/4 inches. Refrigerator waste pipes, except in tenement houses, and all safe-waste pipes, must have brass flap valve on the lower ends. Lead safes must be graded and neatly turned over beveled strips at their edges.

Sec. 28. Vent-pipe Material.—Material for vent pipes shall be of lead, brass, enameled iron or galvanized iron.

Sec. 29. Ventilation of Traps and Soil Lines.—Traps shall be protected from siphonage or air pressure by special vent pipes of a size of not less than the following tables:

Size of pipe	Maximum length in feet	Number of traps vented	
	Mains	Branch	Main vertical
1 1/4-inch vent..........	20 feet	1	
1 1/2-inch vent..........	40 feet	2 or less	
2-inch vent..............	65 feet	10 or less	20 or less
2 1/2-inch vent..........	100 feet	20 or less	40 or less

3-inch vent..............	10 or more stories	60 or less	100 or less

The branch vent shall not be less than the following sizes:

- 1 1/4 inches in diameter for 1 1/4 inch trap.
- 1 1/2 inches in diameter for 1 1/2 inch to 2 1/2 inch trap.
- 2 inches in diameter for 3 inch to 4 inch trap.
- One-half their diameter, for traps 3 inches and over.

Where two or more closets are placed side by side, on a horizontal branch, the branch line shall have a relief extended as a loop. A pipe 2 inches in diameter shall be sufficient as a loop vent for two closets. A pipe 3 inches in diameter shall be sufficient as a relief for three or four closets; and where more than four closets are located on the same branch the relief shall not be less than 4 inches in diameter. All house drains and soil lines on which a water closet is located must have a 4-inch main vent line. Where an additional closet is located in the cellar or basement, and within 10 feet of main soil or vent line, no relief vent will be required for said closet; but where it is more than 10 feet, a 2-inch vent line will be required. Relief vent pipes for water closets must not be less than 2 inches in diameter, for a length of 40 feet, and not less than 3 inches in diameter, for more than 40 feet.

No re-vent from traps under bell traps will be required.

In any building having a sewer connection with a private or public sewer used for bell-trap connections or floor drainage only, a 2-inch relief line must be extended to the roof of the building from rear end of main. House drains, constructed for roof drainage only, will not require a relief vent.

A floor trap for a shower shall be vented, unless located in the cellar or ground floor the paving of which renders the trap [164] inaccessible.

Sec. 30. Horizontal Vent Pipes.—Where rows of fixtures are placed in a line, fitting of not less than 45° to the horizontal must be used on vent lines to prevent filling with rust or condensation; except on brick or tile walls, where it is necessary to channel same for pipes, 90° fittings will be allowed. Trapped vent pipes are strictly prohibited. No vent pipe from the house side of any trap shall connect with the ventilation pipe or with sewer, soil or waste pipe.

Sec. 31. Offset on Vent Lines.—All offsets on vent lines must be made at an angle of not less than 45° to the horizontal, and all lines must be connected at the bottom with a soil or waste pipe, or the drain, in such manner as to prevent the accumulation of rust, scale or condensation.

No sheet metal, brick, or other flue shall be used as a vent pipe.

Sec. 32. Setting of Fixtures.—All fixtures must be set open and free from all enclosing woodwork. Water closets and urinals must not be connected directly or flushed from the water-supply pipes except when flushometer valves are used. Each water closet must be flushed from a separate cistern, the water from which is used for no other purpose, or may be flushed through flushometer valves.

Rubber connection and elbows are not permitted.

Pan, plunger, or hopper closets will not be permitted in any building. No range closet either wet or dry, nor any evaporating system of closets shall be constructed or allowed inside of any building.

A separate building constructed especially for the purpose, must be provided in which such range closets shall be set.

All earthenware traps must have heavy brass floor flange plates, soldered to the lead bends and bolted to the trap flange, and the joint made permanently secure and gas-tight.

In all buildings sewer-connected there must be at least one water closet in each building. There must be a sufficient number of water closets so that there will never be more than 15 people to each water closet.

Separate water closets and toilet rooms must be provided for each sex in buildings used as workshops, office buildings, factories, [165] hotels and all places of public assembly.

In all buildings the water closet and urinal apartments must be ventilated into the outer air by windows opening on the same lot as the building is situated on or by a ventilating skylight placed over each room or apartment where such fixtures are located.

In all buildings the outside partition of any water closet or urinal apartment must be air-tight and extend to the ceiling or be independently ceiled over. When necessary to light such apartments properly the upper part of the partition must be provided with translucent glass. The interior partitions of such apartments must be dwarfed partitions.

In alteration work where it is not practicable to ventilate a closet or urinal apartment by windows or skylight to the outer air, there must be provided a sheet-iron duct extending to the outer air, the area of the duct must be at least 144 square inches for one water closet or urinal, and an additional 72 square inches for each addition closet or urinal added therein.

Sec. 33. Urinals.—All urinals must be constructed of materials impervious to moisture and that will not corrode under the action of urine. The floors and walls of urinal apartments must be lined with similar non-absorbent and non-corrosive material.

The platforms and treads of urinal stalls must be connected independently of the plumbing system, nor can they be connected with any safe-waste pipe.

The copper lining of water closet and urinal cisterns must not be lighter than 12 ounces copper, and must be stamped on lining with maker's name. Where lead is used it must not weigh less than 4 pounds to the square foot. All other materials are prohibited.

Sec. 34. Fixtures Prohibited.—Wooden wash trays, sinks, or bath tubs are prohibited inside buildings. Such fixtures must be constructed of non-absorbent materials. Cement or artificial stone tubs will not be permitted, unless approved by the plumbing inspector and building department.

Yard water closets will not be permitted except as approved by the plumbing inspector and then passed by the building department.

Sec. 35. Privy Vaults and Cesspools.—No privy vault or [166] cesspool for sewage, shall be constructed in any part of the city where a sewer is at all accessible. In parts of the city where no sewer exists privy vaults and cesspools shall not be located within 2 feet of party or street line nor within 20 feet of any building. Before these are constructed application for permission therefore shall be made to the building department.

Sec. 36. Material and Workmanship.—All material used in the work of plumbing and drainage must be of good quality and free from defects. The work must be executed in a thorough and workmanlike manner.

[167]

www.ingramcontent.com/pod-product-compliance
Lightning Source LLC
Chambersburg PA
CBHW031624210526
45464CB00004B/1730